SELLING PEACE

Limit of Liability / Disclaimer of Warranty

The author and publisher of this work have used their best efforts in preparing this manual. The publisher and the author make no representation. or warranties with respect to the accuracy or completeness of the contents of this manual and specifically disclaim any implied warranties of merchantability or fitness for any particular purpose and shall in no event be liable for any loss of profit or any other commercial damage, including but not limited to special, incidental, consequential, and other damages.

Selling Peace - Inside the Soviet Conspiracy That Transformed the U.S. Space Program
ISBN 9781-926592-08-4 - ISSN 1496-6921
©2009 Apogee Books/Jeffrey Manber

All rights reserved under article two of the Berne Copyright Convention (1971). We acknowledge the financial support of the Government of Canada through the Book Publishing Industry Development Program for our publishing activities.
Published by Apogee Books an imprint of Collector's Guide Publishing Inc., Box 62034, Burlington, Ontario, Canada, L7R 4K2
http://www.apogeebooks.com
Cover: Robert Godwin
Printed and bound in Canada

Selling Peace

INSIDE THE SOVIET CONSPIRACY THAT TRANSFORMED THE U.S. SPACE PROGRAM

by

Jeffrey Manber

An Apogee Books Publication

Contents

Introduction	11
Section One: Zero-Gravity Capitalists-1988-1990	17
Chapter 1: Knocking on the Soviet's Door	19
Chapter 2: Perestroika in Orbit	23
Chapter 3: Tricking State and NASA	29
Chapter 4: The World Turns Upside Down	37
Chapter 5: The Mir Really Exists	47
Chapter 6: Aren't Two Routes Better Than One?	61
Section Two: My Boss Semenov-1991-1998	67
Chapter 7: Joining the Space Mafia	69
Chapter 8: NASA Signs with Energia. Maybe.	81
Chapter 9: Propping up the Russian Space Agency	97
Chapter 10: NASA Buys Mir Services	111
Chapter 11: The Competition Virus	121
Chapter 12: With Semenov	127
Chapter 13: 50th Anniversary Bash	139
Chapter 14: Fear Beneath the Petrified Salami	145
Section Three: The Launch of MirCorp: 1999-2000	157
Chapter 15: Rick's Audacious Plan	159
Chapter 16: If You Can't Buy, then Lease	165
Chapter 17: The Merchants of Space	171
Chapter 18: Risky Business	175
Chapter 19: Time to Bottom Fish	179
Chapter 20: But No One Believes In Success	189
Chapter 21: Anderson's Seven Million Dollar Call	199
Chapter 22: NASA's Worst Nightmare	205
Chapter 23: Derechin is Worried	211
Chapter 24: At Least the Communists Are Happy	221
Chapter 25: The Mir is Saved!	233
Chapter 26: Some Get It, Some Don't	237
Section Four: Scheming Politicians, Star-Studded Customers-2000-2001	247
Chapter 27: Conspiracies and Boycotts	249
Chapter 28: The Flyboys of MirCorp	255
Chapter 29: Citizen Explorers	267
Chapter 30: Reality TV Space Race	283
Chapter 31: 2001, Our Space Odyssey	295
Chapter 32: Into the Watery Grave	303
Chapter 33: The Final Confrontation	309
Epilogue	321
Further Reading	325
Index	329

Acknowledgment

There can be no more difficult project than recounting accurately a story involving multiple countries, with dozens of companies and organizations, over two decades and including yourself as a participant. In assuring the accuracy of the manuscript, I am indebted to many old and new acquaintances.

To begin, my thanks to Anthony Arrott for steering me straight on the Payload Systems project, and for providing the pictures of that historic mission. And, of course, for his long-ago offer to visit Moscow. Without that invitation, I doubt I would ever enjoyed the experiences described in this story.

Arnold Aldrich, the NASA veteran who remains a real-life example of the "Right Stuff," kindly reviewed the material concerning NASA's opening to the Soviet Union and corrected several errors.

Other fact checkers included Frank Morring, of *Aviation Week*. Christopher Faranetta had a keen recall of the Energia Ltd. office. Participation in Michael Potter's *Orphans of Apollo* documentary was an opportunity to recall events during MirCorp. The showing of the movie at different locations gave me the chance to become reacquainted with friends, including Tom Cremins of NASA, who provided perspective of how NASA's rank and file viewed Russian cooperation.

John Jacobson and Gus Gardellini went the extra mile in tracking down those wonderful pictures of the early MirCorp experience. Dr. Chirinjeev Kathuria patiently answered my questions on MirCorp's media and Hollywood business. Brian Streidel was also there to review key moments of both Energia Ltd. and MirCorp.

And there were my Russian friends and colleagues, from Moscow industry to the government, who were available to discuss the political forces operating in Russia. It was of satisfaction that all found my analysis of Russian politics to be basically accurate. But in a sad reflection of the times, these three also elected to remain anonymous. My thanks to Nellie Galtseva for her briefing on the linguistic and political significance of the Russian word "Mir." In so doing she gave renewed significance to the hints dropped by Boris Artemov regarding the intent of the Soviet designers of the Mir space station.

Charles Miller gave me insight on the formation of MirCorp. I suppose I was too busy during my time with the company to wonder how it first began. The commercial lease of the Russian space station was at the heart of the MirCorp story, and Jim Dunstan took the time to relive the details of the contract.

Kris Kimel set me straight on the overall tone and direction of the story, and John Mangano and my brother Larry Manber gave much-needed feedback on the title and the intent of the book, as did Martin Wilson, who also designed and maintains the book's web site.

At NASA's History Department, Colin Fries helped in finding the last of the photo's, which meant the path led me to Bill Ingalls and his treasure chest of NASA images.

Perhaps most importantly, a salute to Rob Godwin who when I approached him about writing this story about Russia and the Mir and what I've learned from cooperation, answered "sure, go do it."

The act of writing has been made less lonely by the advent of the personal computer and the wi-fi coffee shop. The folks at St. Elmo's provided me with a noisy, life-filled background for many of the necessary hours in writing. For the cups of coffee and kids running all about, my thanks on breaking the monotony of the writer's lot.

Finally, I'd like to thank some of the many folks not mentioned in the book who have helped me with either space commercialization or on Russian cooperation. A partial list to be sure, but in no particular order: Gary Miglicco, Steven Brody, Peter Nesgos, Tom Whitehead, Perry Streidel, James Asp, Rob Cantor, Vitaly Lopota and Alexander Lopota, Alana Zion, Susan Eisenhower, David Gump, George Koopman, Vinit Nijhawan, Don Brown, Janine Rubitski, Steve Bress and Dan Bress, Frank DiBello, Dennis K., George Berkowski, Francesca Schroeder and Nancy Wood.

Moments before the April, 1961 launch of Yuri Gagarin as the first man into space, an agitated Soviet General Designer Sergei Korolev burst into the mission control room where dozens of engineers sat hunched over the primitive equipment. Korolev began barking out questions. "Tell me the pressure in the fuel lines." "What is the latest weather report? "How is the down range communication?" He came upon a young engineer named Victor Legostaev. "Victor Pavlovich, do you believe in God?"

Some thirty years later Professor Legostaev recounted the awkwardness of the question. "Remember, this was the Soviet era. We were not supposed to believe in God, only in the Communist Party. So his question was a dilemma. But it was a direct question from my General Designer, so I knew I had to answer. 'Sergei Pavlovich, I believe in the space gods.'

"Sergei Pavlovich seemed comforted by my answer and patted me on the shoulder. A few minutes later, he finally shouted out the order that launched the rocket carrying the first man into space."

To those who believe in the gods of space, this book is dedicated.

Introduction

February 21st, 2002

I answered the ringing cell phone, even though it was two in the morning.

On the other end of the line was a semi-hysterical reporter for the AP news service, demanding immediate confirmation whether Britney Spears and Justin Timberlake had booked a ride with my company into outer space.

OK, yeah, sure.

My brain struggled to pull itself out of the sleep induced and in this case, jet lagged fog. I was in Moscow and had fallen asleep with the help of some strong red wine just an hour before. I scrambled to figure the possibilities. A joke? But only a few people had my Moscow number. Trying to buy some time and make reason out of the question, I answered the reporter with my own question.

"What makes you think Britney Spears wants to even go into space?"

Forget it was the curt answer. It's all over the wires.

OK, next try. "Give me your number and I promise to call back in ten minutes. Really, I promise."

Reluctantly the reporter agreed. I switched on the light, threw on some clothes, and called our public relations guru, Jeffrey Lenorovitz in Washington.

Jeffrey answered immediately and confirmed that he was giving out my number to the reporters calling that late afternoon, Washington, DC time. A rumor that Spears and Timberlake wanted to fly into space was going viral and Lenorovitz figured that no matter the late hour in Moscow, better I handle the press.

I turned on the computer and attached my cell phone. A new service in Moscow allowed Internet connections directly from the phone. It was a slow connection and would cost a fortune to track down the story, but money well-spent. The first try didn't go through. Nor the second try. God, sometimes I hated being in Moscow. The third try was successful. Slowly, slowly, I navigated my way through the breaking news stories. And there it was. The source of my lack of sleep was a screaming headline:

Spears and Boyfriend to Become First Lovers in Space?

The story had been broken by the respected French news service "Agence France Press" (AFP) just a few hours earlier. Readers must have been titillated to learn that Lynne Spears, the mother of pop star Britteny (sic) Spears, had angrily slammed down the phone upon being asked about the news, but "did not deny that the couple is negotiating for a space flight" together. So it must be true. Incredible.

The French reporter broke with the sensational news that female pop star Britney Spears was in negotiations to fly to outer space, along with her boyfriend Justin Timberlake. These plans came on the heels of the rumors that Justin Timberlake's bass singer in the boy band 'N Sync, Lance Bass, also wanted to experience a ride into the new frontier of outer space.

Revealed the reporter, "Spears, like Bass, is reputedly negotiating with Netherlands-based company MirCorp, which has an agreement with Russian spacecraft manufacturer Energia to market tourist flights.

"In 2001 millionaire businessman Dennis Tito became the first "space tourist" when he flew to the International Space Station aboard a Russian Soyuz taxi flight. Tito's trip prompted a bitter dispute with NASA.

"South African businessman Mark Shuttleworth is scheduled to become the second space tourist in April, following the establishment of new rules about who is qualified to fly to the space station. But if Spears and Timberlake actually fly, they will not visit the space station and NASA will have no say on their activities."

The article then provided some history of Hollywood's interest in space, saying, in part, how "This seeming rash of interest in spaceflight among musicians is actually not new. Folk singer John Denver, killed in a 1997 plane crash, expressed an interest in flying to the now defunct space station Mir in the early 1990s."

The mention of legendary folk singer Denver struck me as a surprising inclusion, even when reading the article in the dead of night. Very few people knew of Denver's serious attempt to journey to space.

I still view the moment as an opportunity that slipped away. Victor Legostaev, the senior official in charge of international business development at the giant Russian space company Energia, had summoned me to his office and explained in his bass voice that an "American man with guitar wants to fly with us. The problem with all men with guitars," philosophized down-to-earth Legostaev, "is that these sort of men never have money. But he wants to fly and write a song for us." I asked who the man was. The answer was John Denver. I would later learn that Denver had tried to fly with NASA in the mid-1980's and had been refused. The composer of "Rocky Mountain High" and other mega hits then went to the Soviets, who offered to fly the musician for, of course, a fee. For the past two years, Denver had been struggling to find a way to realize his dreams, but without paying cash up front.

Not for the last time did Legostaev, a Hero of the Soviet Union, and I fight over differing philosophical outlooks on the new business of space tourism. For me, the advantages of sending John Denver into space were clear, at whatever cost. The publicity, the opportunities for branding and the resultant royalties resulting from launching a mega-entertainer would have started a new sort of entertainment gold rush. What a promotional and entertainment coup that would have been. The more conservative Legostaev objected to what he saw as taking a financial risk with a folk singer. I would think back on this missed opportunity later, after Denver was killed in his experimental plane. What great tear-jerking songs he would have written while gazing down on the blue orb that is our earth, and how often those songs would have been played during sentimental times and long lonely journeys.

The Russians were willing to embrace our capitalist business models, but still required cash up front. This hesitation by the top Russian space officials to share in expected revenue remains dominant, and it is a shame, as it has allowed some wonderful media and entertainment opportunities to remain out of reach.

Then the breaking French news report got to the really good stuff, when the reporter wrote how "The Russian spacecraft has two compartments that can be sealed with a hatch. The rumors that Spears and Timberlake hope to fly together has prompted speculation among her fans, and criticism from her detractors, that the two are planning a "zero-gee love tryst" in the words of one critic. Shelly Travers, head of the non-religious morality group League of Decency, who has previously criticized Spears, said "Now it looks like she is going to take lust to new heights."

Space expert John Pike, of GlobalSecurity.Org, a think tank in Washington, DC, was quoted as suggesting, "It used to be that it was the superpowers competing to put people in space. Now it is apparently rock stars." As for the rumors that the couple may be planning a romantic rendezvous in orbit, Pike joked "NASA has spent a lot of money studying long-duration spaceflight. It is a long way to Mars. Maybe they will learn something useful about how to pass the time." Pike is always good for a quote, and he must have had a fun time with this one.

All night my phone continued to ring. Reporters, editors, producers, were all demanding a quote confirming the story. As president of MirCorp, the company named in the above scoop, I was caught off-guard. We knew nothing about Britney Spears or Justin Timberlake wanting to fly together into space. But the story was being carried by a respected news organization, and much of the rest of the other information seemed accurate. We were trying to arrange for Lance Bass of 'N Sync to fly to the space station, and the story accurately captured both NASA's opposition to Dennis Tito and the space agency's new neutral position on flying Shuttleworth, the South African co-founder of the Internet payment system Paypal.

We had first learned about Bass through a news story, when the young 'N Sync singer told a packed Los Angeles press conference of his dream to fly into space, and his willingness to undertake a television show to pay for the seat, so why not Britney?

Three in the morning. Four in the morning. Five in the Moscow morning. Over and over I repeated to the reporters that we knew nothing. Some wanted to know the history of sex in space. Whether we had spoken with Britney, whether sex would be permitted between the two pop stars. And how much the flight would cost the stars. Twenty million? Forty million dollars?

One of the entertainment producers gushed that the event, if true, would be a bigger media crunch than "the Beatles invasion." As I scrambled to understand the situation, I also sought to engage the reporters on an overlooked facet of the breaking news. Why was MirCorp working with Russians, and not with NASA? Put another way, why did Lance Bass have to come to Russia? Or John Denver? Or Radio Shack, which was filming a commercial on the Russian side of the new international space station. And now maybe Justin Timberlake and Britney Spears. To all the reporters I wondered aloud just why did the United States balk at allowing traditional American commercial activities, whether advertising or entertainment into the frontier of outer space.

Just how has it come about that we owe it to the Russians for showing that capitalism and tourists can thrive, like dogs, monkeys, yeast cells and fighter jocks in the zero-gravity of the space station? My ruminations fell on deaf ears. All during that snowy February night the reporters, the researchers, the producers, just wanted to know if there would finally be sex in space. And by morning, I was curious myself.

The Britney Spears story faded away in a day or so. I never learned where it came from, but was forced to spend several long meetings with my Russian colleagues explaining to them what the fuss was all about. Even the head of Energia, Yuri Semenov, called me into his office with Victor Legostaev to discuss Timberlake, Spears and Lance Bass. The whole situation was frustrating, since no reporter had taken my bait that frantic evening, meaning that the paradox of Americans trekking to Russia as a commercial gateway to the space frontier remained unexamined by our media and more importantly, by us.

It is a discussion that we must have.

America has spent over $100 billion dollars building a space station that more than likely we did not need. We have spent an equal sum on a transportation system so flawed that the weak links were identified by journalists before the first launch. Not to mention two tragic accidents. Even today, $200 billion is a large sum to have wasted. We clung far too long to a business model that was a relic from the Cold War. Frozen out of our space program were innovative concepts and projects from our own industries and from those in foreign markets.

Sure it's fun to think about Britney Spears or Timberlake having sex in space, but it's also critical that we develop the right mix of government oversight and private initiative that allows space exploration to be an extension of who we are. Just as the Internet has unleashed a powerful explosion of democratic values through social networking, decentralization of news and elimination of geographical boundaries, so will private sector space exploration cause a similar social disruption.

The good news is that we seem to have figured it out. Private companies today are developing innovative space projects with the assistance and support of NASA. But forgotten is the fact that this industry grew so frustrated that it took entrepreneurial Americans and powerful Russians working side by side to launch American style capitalism into outer space.

How we got to this point is a hell of a story, filled with all sorts of colorful, stubborn and very smart characters. None more unusual than Energia's head, Yuri Semenov, who for much of the 1990's was my boss. This bellicose and difficult Russian saw first, before NASA, the value of using the free markets to support manned programs. The other major character in this story is the Mir space station itself.

The Mir was spawned in the Cold War by the secretive Soviets, morphing in the late 1980's into a symbol of perestroika, then reincarnated as a platform for Russian economic reform. Next it became a run-down mockery of all that was then wrong with the Russian society of the mid-1990's, but for a brief tantalizing moment, right near the end of the decade, the Mir became a beacon on the path to commercial markets and competition in outer space. It is this final reincarnation of Mir that I choose to remember.

By a series of circumstances I was present at many of the key milestones, as Americans came to grips with the paradox of having to journey through Russia to find a way to the space frontier. This is my recounting of what I saw, filled no doubt with my own biases and opinions.

Let's start at the very beginning, when a Boston company took the first steps to circumvent the NASA monopoly on manned exploitation of space. Why the company took that fateful step, and the consequences that rippled out across the American political landscape, is the tale of this book.

Section One: Zero-Gravity Capitalists
1988-1990

Space is just another place to do business

-U.S. Secretary of Commerce Malcolm Baldrige

Chapter 1: Knocking on the Soviet's Door

In early spring of 1986 two friends of mine walked out of a historic stone building on 16th street in Washington, DC. It was one of those buildings that dot much of the downtown Washington landscape. A curved driveway, an ornate gate, neat rows of flowers maintained by a professional gardener, and very often, as in this case, no inscription to publicize the nature of the business taking place inside the stone walls. Turning left on 16th Street, Byron Lichtenberg and George Economy walked towards Lafayette Park and the White House. Within two blocks the street ended at the Park. Unsure which way to turn, George checked his map. Still unsure, he then asked a woman behind them directions to the nearest Metro for returning to National Airport.

"I'll tell you how to get to the Metro," replied the woman, showing her FBI credentials, "if you tell me what you were doing in the Soviet embassy."

Lichtenberg and Economy had come to Washington that day solely for a clandestine meeting with the scientific staff at the Soviet embassy. As two of the senior officers of a small research company located in Boston, they were frustrated with NASA's continued reliance on the fragile space shuttle. They were also frustrated with how NASA discouraged private sector participation in space, despite all sorts of promises to the contrary. Befitting the analytical mindset of the company founders, a remarkable conclusion had been reached. If NASA was blocking the company's plans to develop in space, then it was time to seek another route. That logically meant contacting the Soviets to explore whether the one-year old Russian space station Mir was available for commercial research.

Payload Systems was formed in 1984 after Dr. Byron Lichtenberg, the boyishly handsome professor at the Massachusetts Institute of Technology, became the first non-government passenger to fly into space on the space shuttle. Byron worked on a West German research mission known as the D-1 lab mission. He was intrigued by the possibilities of commercial space research and a year later Lichtenberg formed Payload Systems, with Dr. Anthony Arrott, who had worked with Lichtenberg at MIT in preparing research hardware for his shuttle mission.

The co-founders believed there was an opportunity to serve those researchers and scientists interested in basic research and manufacturing in space. "NASA equipment was 10 years old," Arrott told me at the time, "and the bureaucracy was already difficult for industry researchers." The equipment was so old because it had been built assuming the space shuttle would be ready in the mid-seventies. But it was not. It was late, and researchers were left with last generation electronics and materials.

Of interest to entrepreneurs like Arrott and Lichtenberg were the discussions taking place within the free-market Reagan administration to encourage the underwriting of space research by non-NASA organizations and researchers. Let space research, for example, be conducted by the National Science Foundation; which is tasked to coordinate our government's support for scientific research. Those who believed in the Reagan philosophy of less government were quietly looking to restructure NASA to reduce its power. For die-hard Reagan believers, NASA represented everything that was wrong with the federal bureaucracy-a cumbersome and expensive government agency blocking American ingenuity.

The founders of Payload Systems saw a new industry developing and were ready to take part. The Wellesley company, located in a funky warehouse area of the changing suburb of Boston, developed a solid business. Their services ranged from designing equipment for use in space to planning shuttle-based research programs. This sort of private participation in the shuttle was a novel idea for the space program, a privatization of what had been a NASA monopoly: designing, developing and operating space research projects.

The client list quickly included large aerospace firms and the government agencies of France, Germany, Japan and Canada. The staff grew to some twenty space engineers and researchers and in the mid-1980's it was the largest company in the world devoted solely to space research and hardware. What had everyone so intrigued was the idea of using the microgravity of outer space for developing strange new alloys and even pharmaceutical drugs. The example always used on novices like me was to explain how oil and vinegar doesn't mix in our salads here on earth, but does in space. Apparently with a bit of technical wizardry you can get this sort of new compound to "stick" in the zero-gravity of space and one would have a new drug or stronger alloy. Researchers can also grow larger drug crystals in orbit and learn all sorts of things.

NASA touted microgravity research as the wave of the future, leading to innovations including (literally) the cure for cancer and a new generation of computer chips. It was the ultimate payback for the billions invested in the space program, an entirely new research industry, and NASA was determined to serve as the gatekeeper.

After the Challenger accident, the impetus for Payload Systems to use Soviet space assets made practical business sense. The heavy marketing of the space shuttle program to non-aerospace companies had attracted all sorts of forward-thinking companies, but pharmaceutical companies were now becoming turned off by the mechanics of working with NASA. The government bureaucracy, the red-tape, the erratic flight schedule, all made commercial space research virtually impossible.

The grounding of the space shuttle fleet left many companies, like 3M and Johnson and Johnson, even tractor maker John Deere, with space research departments and no access to the zero-gravity of space. It was a time of intense frustration. NASA had preached with evangelical zeal a new era of space-age miracles, and then the door was shut on the Promised Land. Program managers and industrial researchers faced a stark choice: shut down their space research programs or find another way to get to space. Most chose the former, but a few sought out Payload Systems to brainstorm whether there was another practical solution.

Just a year earlier the Soviets had launched their latest space station, called Mir. Russian for "peace", the space station had been ridiculed as nothing more than a tin can by American supporters of our planned space station Freedom. But this tin can had docking ports, and slowly, over the next couple of years, modules were added and soon the lowly tin can was a good-sized home orbiting through the dark void of space. At about the same time the new Soviet premier, Mikhail Gorbachev, came to power proclaiming a time of change, or perestroika. All government programs, whether aviation or the ballet or outer space, had to exist on commercial funds. It was a stunning evolution to a Soviet centralized economic system that had never trusted free markets.

NASA officials completely ignored the historic developments taking place in Russia. No one from the space agency suggested, for example, that during the time the Shuttle fleet was grounded it might be a good idea to have American industry use Soviet space assets. So the guys at Payload Systems, on their own, without any government involvement or approval, decided to simply knock on the door of the Soviet embassy and ask if the new openness, or glasnost as it was called, included allowing pharmaceutical researchers to undertake months long experiments for drug research on the Soviet space station.

The answer was an emphatic yes.

Would they be open to a set price along commercial terms?

Once again, the answer was yes.

And finally, would the Soviets be willing to keep the results secret and the terms and conditions secret, to allow Payload Systems a competitive advantage?

Yes and yes.

It was this audacious move by businessmen, stymied by the situation with NASA, that opened up a new pathway for space exploration. Like the dream of the 15th century explorers, seeking a shortcut to the land of spices, this northern passageway also would take us through Russia.

Chapter 2: Perestroika in Orbit

I was stunned at the audacity and simplicity of the idea when Arrott came by and spilled the beans. Not only was NASA now grounded by the Challenger tragedy, but so too the ambitions of many entrepreneurs and companies to fly space manufacturing projects. No one had thought to knock on the door of the Soviet embassy and politely ask about flying on the Mir. No one except Payload Systems.

I listened in disbelief as Arrott described his dealing with the Soviets. By now they had met several times. He even had a family friend in Moscow doing the necessary direct negotiations. Arrott spoke of how the Russian officials were business-like in those negotiations. How they had agreed to a set price, which would be kept secret. This was impossible with NASA.

The Soviets had agreed to keep all issues of intellectual property rights with the customer. This was also then impossible with NASA. They had also agreed to work with Payload Systems for two flights of two months each, in order to allow plenty of time for the pharmaceutical crystals to grow. This was technically impossible with NASA, given that space shuttle flights were about a week in duration.

Anthony Arrott's embrace of the Soviet marketing effort forever changed the American space program

Photo: Jeffrey Manber

Arrott spoke about working with three organizations. One was the very public Glavkosmos, a Gorbachev organization set up to attract commercial space business. Assisting Glavkosmos on the business side was an organization known as Licensintorg, which for decades had done business with America. It must mean that Licensintorg had the necessary bank accounts and experienced contract negotiators. Later I would learn that it too had been given rights by the Soviet leadership to market satellites and space services. And finally, there was a more mysterious organization called NPO Energia that somehow controlled the space station. This was weird news. Didn't the Soviet government control the space station? Even then, the very mention of Energia caught my attention. In terms of heritage, accomplishments and all-stars, they were to space what the Yankees were to baseball, without a doubt.

Their first leader was Sergei Korolev, who was the founder of the Russian space program. It was under his guidance that the Russians beat America to space by launching Sputnik, the world's first satellite. Korolev also implemented the flight of Yuri Gagarin, the first human to venture into space. Korolev's life was kept a secret by the Kremlin until the General Designer suddenly died on an operating table in 1966. Only then did the world learn how this survivor of Stalin's prison camps emerged to develop Russia's rocket program using sheer willpower. Pictures of him show a dark, brooding and powerfully built man often dressed in black.

The name of his organization reflected the zigzag course of the Cold war. During the 1940's and 1950's it went simply by the name of "Department 88". Very cool. Very mysterious. Then, just before Sputnik was launched it became known as "OKB-1." Still pretty cool, like one of the robots from the Star Wars movies. In the mid-1970's OKB-1 developed into NPO Energia.

The NPO stands for Scientific and Technical Organization. The "Energia" means Energy, and it was a simple ruse to disguise the precise activities of the space organization. Maybe the name change was done because their engineers and cosmonauts took part in the Apollo-Soyuz program, and they knew it was a little too secret-agentish to introduce themselves to NASA engineers as "OKB-1 at your service". Energia is a nice non-descript catch-all name. Though even when working with NASA in the 1970's it stayed secret, and Energia engineers traveled to Houston as members of a scientific academy. It was NPO Energia's engineers that were also meeting with Payload Systems.

Hearing his story, one of my initial concerns was Anthony himself. Of medium build, his piercing blue eyes signaled a hyperactive businessman who loved the Machiavellian maneuverings of business. Often, board members of his company were pitted against one another and arguments waged long after a decision had supposedly been made.

Nonetheless, even allowing for some hyperbole, listening to Arrott was beyond anything I had heard before. He was pulling back the curtain on the Soviet Union's deepest industrial secrets. How they were changing under the new reforms and what their true intentions were. This was not a caviar or vodka deal. This was the heart of the Soviet Union's military establishment. NPO Energia, or Department 88, developed the Soviet ICBM missiles. Imagine, working with them on a commercial project.

I just shook my head.

Arrott didn't tell me about his confidential plans because we were friends. No, it was because I had recently begun working in the U.S. Department of Commerce. I had been enlisted by a zealot critic of NASA named Gregg Fawkes, who was most proud of his direct heritage to the 16th century Roman Catholic British rebel Guy Fawkes. With his tall lean frame and handlebar mustache, he could easily pass for his namesake.

At the time I was pretty burnt out. Years of writing about the frustrations of the fledgling space industry had led me to take up an offer from Jim Samuels, a veteran aerospace analyst, to help run a space fund out of Wall Street. But just a year later the entire department had been booted out of Shearson Lehman, the huge financial Wall Street financial company. Ghoulish as it may be to say, the Challenger accident was actually good news for the private launch vehicle industry, which was one of the investment targets of our department. That fact was lost on management which, seeing the tragic news on the television, and learning of the grounding of America's space shuttle fleet, eliminated the department. Maybe the fact we were losing every dime we had invested, given how badly the commercial space industry was progressing, also played a role, but I know we would have soon turned the corner. Really.

One of the new generation of space entrepreneurs butting heads with NASA was a real clean-cut, well-dressed Texan named Russ Ramsland, then head of a company called Microgravity Research Associates. Russ was from Midland, Texas, friend of the Bush family, and he had come to believe that if nothing else, NASA had to be cut down in size given its reluctance to work with the private sector. In this regard he was part of an influential group of Reaganites who were determined to break the grip NASA had on space policy, industry regulations, and civil space operations.

I had interviewed Russ several times about his efforts to grow gallium arsenide chips in space, for high-speed computers, and he apparently recommended me for a new initiative in the Administration, called the Office of Space Commerce. The department would be in the Office of the Secretary, in order to shield it from NASA supporters on the Hill. Now, this is Washington-speak here, but there was no way that Congressional space program supporters would allow industry to have a voice on space in the government that was not through NASA. This may sound strange, but that's the way Washington then operated. So to circumvent the opposition, we were set up informally within the office of Secretary Malcolm Baldrige. This sort of intrigue appealed to me, and I agreed to come down to Washington from New York for a one year period.

My first day at Commerce was memorable. First I was fingerprinted. Nice to know you can be fingerprinted without having been arrested. Then I took an oath of allegiance to the Government, which woke me up to the fact that this was not a normal job. And finally I was escorted by a security guard to the empty Space Commerce office. It had several desks, a small American flag and no computers. We were brand new, with a few recruits from elsewhere in Commerce. Then in the afternoon Gregg Fawkes took me to meet his boss, a die-hard right-wing former Marine named Bob Brumley.

Brumley served as the General Counsel of the Commerce Department, meaning the top lawyer and overall number three in the building. His office was considered part of the Secretary's, and so it had (like mine would) the blue carpet indicating as such. Silly stuff, but we learned quickly to use the blue carpet to cut corners on ordering computers, travel requests and that sort of thing. It was Brumley that had maneuvered to establish the new Office. The meeting with Brumley was short. Really short.

I know you have worked with NASA and the space industry, he began. He cut me off with a wave of his hand as I began to protest that I had never worked for NASA.

I know you have a lot of friends there.

So you will be the good guy. We need you to smile and agree to work on all sorts of projects. Gregg and I will be the bad guys. We will lie and cheat and do what we can to screw NASA. We've had enough of NASA's crap. It's time to break-up the bastards.

OK, got it. Simple enough. Yes boss.

Both Arrott and I knew it would be an uphill struggle to get the Administration's approval. To work with the Soviet Union an export license was required. The question was which agencies had regulatory responsibility. Standard procedure would be to file the license request with the State Department, which had all sorts of experts on missiles and rockets. The chief function of their role seemed to be to maintain the status quo against the Soviets, unless directed otherwise at the highest level, meaning by the president himself. Two red-button terms, the Soviets and space, would suggest that this was an issue for the State Department.

I was new to Washington but not to the space industry. I clearly understood that State had no desire to see cooperation develop, even if it helped American industry. The request for a license to work with the Soviet space station Mir would be dead on arrival. And if State reviewed an application for cooperation in space, it would routinely be sent to NASA for review. A second death, if that was possible.

To make our chance of success even grimmer, recent applications by U.S. satellite manufacturers submitted to State to work with the Russian Proton launcher had all been refused. Arrott knew this. What he was banking on was some roundabout strategy that would get the license issued while staying clear of knee-jerk opposition. Probably, without help from inside the Administration, this guerrilla style operation would never be implemented. Would Commerce help?

With a great deal of trepidation I took the opportunity to Fawkes. The poor man was completely overworked-we were a small and young office fighting all sorts of policy battles. Trying to push through an interim space station instead of the NASA palace in the sky called Freedom was one initiative. The interim station was the work of two American space legends, Astronaut Joe Allen and Dr. Max Faget, whose name was on the patent for the Mercury capsule. Both were concerned about committing America to the multi-billion dollar station before testing the waters with a smaller version.

We were also working with an entrepreneurial venture called PanAmSat that sought to break up the international monopoly on telephone and video, via satellites, controlled by the international organization called Intelsat. And now along came a request from a politically unconnected company to help them work with the Soviet Union, and take on the State Department, NASA and Congressional supporters of NASA.

Fawkes decided to throw the dice and leave it up to Bob Brumley. At the appointed time a few weeks after my meeting with Arrott, we walked together to Brumley's office. As he was of course running late, Brumley's secretary let us into the spacious office. Potted tall plants, high windows facing towards the Washington monument, pictures of Brumley with the Secretary and with the President. It was like something out of a movie. Fawkes nervously paced up and down. A chance to help Payload Systems obtain the export license was the very sort of reason he had signed up with the Office of Space Commerce. But Fawkes quietly voiced to me the same worry about Brumley that I had about him: would it be too radioactive for either Brumley or the Secretary?

So we waited for the chief legal officer. The door swung open. "OK, what are you guys up to? Getting us in trouble, I hope?" Boomed a cheerful Brumley as he marched into his office and took in the nervous Fawkes.

Fawkes let me do the explaining.

I laid out the situation as concisely and quickly as I could. As Brumley's time was usually booked weeks in advance, an interruption from the lowest White House official or from the Secretary could easily freeze the issue for another frustrating week. So I ticked off the facts.

An American company whose clients were stranded by NASA. Had real commercial business with solid commercial research clients from the pharmaceutical industry and needs access to long-duration flights. Nothing in America for years to come. Solution? Work with the Soviets. Brumley was intrigued, but not yet sold. He harrumphed something like "no chance the Soviets will play ball with an American company." But I was rolling, building up steam. I told him about the clandestine meetings. Discussions in Moscow. About the commercial terms. Protection of intellectual property. A set price.

"How much?" this free-market disciple of Ronald Reagan immediately demanded. Brumley was instantly curious, like I was, as to the price to fly on the Soviet space station. Was it cheaper than NASA? More expensive? There had never been competition in manned space. What was the price required by the Communists?

For the first time, I got to shoot Brumley my best condescending look. "It's a commercial secret."

Brumley broke into the biggest shit-eating grin I'm sure anyone has ever seen in the offices of the senior staff of the Secretary of Commerce. Then he started to laugh. Fawkes quickly transformed into one happy overworked policy wonk. This was the battle we had been waiting for. We would figure out a way, some way, to let U.S. industry commercially use the Soviet space station, and to do so without the involvement of NASA.

Chapter 3: Tricking State and NASA

The deal with the Soviets became our number one issue. The goal was to get Payload Systems the necessary export license. Somehow. Could we do it?

Brumley's first task was to brief Secretary Verity, who had joined only the month before, after the sudden death in a freak horse riding accident of Secretary Baldrige. If Brumley received the Secretary's permission, he would brief his colleague at the Pentagon, and some friends at the White House and see just how the path could be smoothed over. At the same time we would reach out to export colleagues at Defense and Commerce. Even within our building, we had to tread carefully. It was a delicate dance. One could not just demand the export office sign off on a controversial deal. That sort of political pressure "from above" only takes one disgruntled guy and the story would be on the front page of the *Washington Post*. But the export officials would also be unlikely to approve unless there was a quiet green light from above. Assigned to handle this very delicate task was Hank Mittman, the chain-smoking institutional memory of our young office. In his early 60's, Hank had already retired from industry and served almost a

Ronald Reagan's free market philosophy was embraced by the Russian people

Photo: Ronald Reagan presidential library

decade in the Commerce building. He knew the minefields of export regulations inside and out-and had been recruited by Fawkes to handle industry issues regarding Reagan's planned space station Freedom.

It was a tall request we were asking of Secretary Bill Verity. Consider his position. Verity was a Midwestern businessman, former chairman of the giant AMCO steel conglomerate. At 69 years of age, he probably regarded the appointment as vindication of a long and successful career and an easy two-year assignment until the new administration picked his successor.

However, maybe there was more to him than I realized. Reagan's appointment of Verity had sparked mutterings from the far-right that distrusted his international perspective and willingness to work with the Soviet Union. An issue was made of how Verity had denounced the Jackson-Vanik Amendment, which made U.S. trade with the Soviets dependent on the Kremlin's willingness to allow Jewish emigration. At his confirmation hearing, Verity called increased trade with the Soviet Union a "vital need" of the United States. So his approval would seem more likely, unless he didn't want to be branded a Russian-lover in the first weeks of his tenure. Also, given the stealth strategy we were employing, would he want to become known as the guy who snuck around State to work with the Soviets?

One of my first lessons of Washington was hitting me over the head. That the true issues of a policy battle were sometimes not the issues on the table. What I mean is, that for some in Commerce this was a matter of winning against NASA. Others drew a larger circle and viewed this as a chance for trade with the Soviets, and some drew an even larger circle and saw our quest as another chance to slay the Evil Empire, and teach them the moral righteousness of American-style capitalism. Under the new Secretary of Commerce, my working night and day to get the export license might unfortunately result in a signal from the Administration that it was time to untie Jewish emigration from trade treaties with the Soviets. So unless you completely control the situation, you have no idea what the hell is going on.

The Secretary listened politely to Brumley and promised an answer soon, meaning that the effort to allow Payload Systems to work with the Soviet space station now hung on which version of Secretary Verity would make the decision. Bill Verity as the believer in trade with the Soviet Union, or it could be instead Bill Verity as the humble new junior team player of the Administration. We didn't have much time, as the rocket carrying the Payload Systems research project would blast off in one year. In the space business, one year is not a long time to develop the hardware, have it tested, have the research decided, get the payload to Moscow, have them approve everything and then have it transported to the launch site.

Hank Mittman swung into action, not wanting to wait for an answer from up above. Mittman had grown sick of hearing my "two-Verity" theory and decided time was indeed wasting. The veteran survivor of countless Washington battles arranged a meeting with export officials from the Department of Defense to take place in the cafeteria of Commerce. There was an interagency meeting on an export dispute taking place in the building and most of the key officials would be there. When they broke for lunch Mittman would separate the representatives from DoD and Commerce's Export Control Administration from the larger group. I was to walk through the cafeteria and Hank would coincidentally call me over to join them. There was no more graphic depiction, that the enemy in this policy battle was not the Soviets, but other members of our Administration.

Once Hank called me over we would then have about twenty minutes to explain the export license application, what it was, what it wasn't, and answer any questions. Hank would do the rest. The advantage of the serendipitous get together was that there would be no record of a meeting. If the license blew up in our face, no one at NASA or Congress or a reporter could find proof of an interagency conspiracy.

The bag of tricks known to Mittman never ceased to amaze. This veteran took the time to teach me the dangerous phrases in agency agreements that, as he would say in his tobacco induced gravelly voice, with his blue eyes twinkling, "you could drive a truck through." Phrases such as "or as appropriate," "all available means" were more than likely minefields to be struck out of any proposed contract or agreement. I don't know how many times his sage advice saved me in the years to come.

The five men sitting around the cafeteria table were all, like Hank, in their late 50's or early 60's, with one exception. Their faces were open, telling me they were pre-disposed to listen to the facts. Their questions were specific. How could the project be packaged? How much would the Soviets know about the mechanisms involved? The type of pharmaceutical crystals? Who on the Soviet side would receive the money? Why not conduct the deal through NASA? Maybe for Brumley and Fawkes it was a policy game, but these guys were taking their role in export control seriously. I left the cafeteria feeling nervous.

A few days later Fawkes delivered good news. Verity had approved Commerce supporting the license, as long as our export department was comfortable. So Verity had stayed true to his business beliefs and did not shrink from taking on a controversial issue right after joining Commerce.

The Secretary did not overtly endorse bypassing State. That would be left on a lower level, meaning Brumley, so it was his ass on the line if there were problems. When he gave his go-ahead, the Secretary also expressed his hope we could keep him out of trouble.

I was really sorry when we were unable to satisfy that simple request.

Payload Systems didn't have a lot of money to burn, but the company had hired the high-octane law practice of Hogan & Hartson. This is one of Washington's powerhouse firms. The willingness of the partners to take on the case suggested that they viewed trade relations with Russia as a growth area, if the Gorbachev reforms stayed on track. The lawyer handling the case was Ann Flowers. In her mid-thirties and still single, Ann lived for three things: her legal cases, her cherished West Virginia Mountaineers football team, and charity work. She cut no corners, volunteering frequently in soup kitchens in Washington, DC, especially around the holiday time.

Flowers had been researching previous cases that had failed at State. She had also thrown herself into whether drug research was on the munitions list. The Munitions List is a horrid black hole that permits mid-level bureaucrats to wreak havoc with international trade policies that would benefit this nation while also causing a loss of power for their agency.

Yeah, I'm biased.

It is better known today as the ITAR list, and it is the regulations concerning mostly military and aerospace goods and services that cannot be exported to the Soviet Union without the approval of the Department of State and the Pentagon. If pharmaceutical crystal research was not on the list, then the licensing authority would rest with Commerce, with a second opinion required from the Department of Defense. But all of this was a very gray area, given that there was some technology transfer from American researchers to the Soviets. Amazingly, we would come to learn there was also no sort of blanket regulation requiring State Department jurisdiction on commercial services aboard Soviet space hardware. This was a major mistake in government control that was later corrected. That's the way government works, they are good at closing holes after the fact.

Weeks of secret meetings dragged on.

Ann and Hank were in quiet discussions with the export officials at Commerce and Defense on what sort of restrictions would make them feel comfortable enough to grant the export license. The Pentagon export officials were insisting on zero technology transfer, especially given the heat that was anticipated once NASA learned of the project. The engineers at

Payload Systems developed a simple black box that would contain the experiment. Inside the block box were four vials containing the pharmaceutical crystals. At a certain time, the cosmonauts on the space station Mir would turn each of four screws, which would mix different components and commence the growing of the crystals. Two months later, at the proper time, the screws would be turned again by the cosmonauts, stopping the growth of the crystal. No transfer of technology from the hardware.

Sounds simple, but each new change in the hardware required coordination between the researchers from the pharmaceutical companies, the engineers, the export officials and finally, the Russians. We were also concerned as to whether the Soviets would permit a sealed box to be loaded first onto their cargo ship and then unpacked onto the space station without knowing the exact contents. NASA never would and never did. To our pleasant surprise all the Russians required was a breakdown of the class of the chemical compounds and a statement regarding the hazards. In other words, they trusted the foreigners not to put something dangerous onboard their manned space station. And to honestly stipulate the class of compounds.

The same held true with the other required supplies and components. The researchers wanted a tape for the video camera already onboard the station, in order to record what the cosmonauts did, in case something went wrong. DoD officials thought this a great idea and insisted it be included as a requirement. The video tape for use on the space station was purchased at a local drugstore. Again, completely impossible with NASA, which insisted on subjecting every bit of cargo for space shuttle to chemical analysis.

I learned in bits and pieces more about how Anthony was handling negotiations. His father, Professor Anthony S. Arrott, was a respected scientist who had long-standing professional contacts within the Soviet scientific community. This helped tremendously as there was less chance that the proposed project would be viewed as a political stunt, something feared by the Soviet side. His sister was with the Voice of America and also had inroads. There was a family friend doing the face-to-face meetings, relaying the changing requirements from both sides in real time. It was probably the best way to conduct the business, as it reduced the chance of any leaks.

There was one more parallel undertaking required of our stretched and weary office. That was to assist in the necessary legal determination by both Department of Defense and Commerce that the license need not be shown to the State Department. Everyone in the export departments wanted their actions protected, and that required a written determination by the lawyers. It seemed clear there would be no technology transfer. It seemed clear that

Commerce could approve because it would help American industry, and there was no such facility in the United States. But shouldn't State and NASA review?

This is where the politics really got involved. A legal determination was drafted on the most fundamentalist reading of the regulations. Years of precedence and common courtesy were ignored, but not before there were heated arguments and tension filled meetings. If the license generated political heat, one had to make sure there were no sloppy mistakes. Or no mistakes at all. Everyone was on edge. Easily Brumley could be hung out to dry and Fawkes as well. So too the experts in the Export Department. Yet everyone believed the project was inherently beneficial and worth the risks.

In the end each agency signed off for differing reasons. The guys over at the Pentagon welcomed the license to get a calibration on the validity of the Gorbachev reforms. Were Soviet hardliners really willing to let American researchers work on their space station for a commercial fee? The Pentagon officials also quietly agreed to not consult with their colleagues at the State Department as a means of asserting their own strength on foreign policy. It was an easy decision to keep the application from NASA. Pentagon planners remained frustrated by how NASA had long insisted that all DoD satellites launch on the space shuttle, even after it was clear the shuttle would not live up to expectations and military satellites sat grounded. We quickly picked up critical allies all over the Administration, including those at Treasury, Commerce, Transportation, and in the White House who, along with the President, welcomed Gorbachev's embrace of private markets. For many true believers the collapse of their command economy was the final confirmation that Reagan and America would win the Cold War.

But right now Brumley's motivation at Commerce was nice and simple. Here was a holiday treat to screw NASA in favor of industry.

As everything was coming together for us, a new data point strangely materialized. In early December Soviet premier Mikhail Gorbachev came to Washington to meet with the President. His trip was a media sensation. The Soviet leader plunged into the crowds, shaking hands, signing autographs. Unlike earlier Russian leaders he was a hit with the American public.

I was walking down the "blue carpet" corridor when I saw one of the older assistants to the Secretary, a man brought in by Verity to work with him. He came up to me, and after saying some nice words about my work, he dropped a bombshell. Verity had met with Gorbachev just the day before. I had heard nothing about this. The subject was trade and new markets. Had they discussed the Payload deal? I didn't ask. The whole conversation made me uncomfortable. I mean, part of me was thrilled. Here I was, now 32 years old, brand-new to Washington, and now maybe the Secretary of Commerce and the leader of the Soviet Union had discussed my first policy battle. But what would it mean if they did? I had thought we were taking advantage of the new Secretary. I was changing my mind.

Fawkes and I had begun to consider what should happen if the license was granted. Stay quiet until it leaked out? Hold a press conference? Have Payload Systems hold a press conference? Arrott was pushing for maximum publicity to attract more customers. Brumley concurred, as he wanted to let the political community know that Commerce was a major player, even at the expense of others in the Administration. In early December I approached Fawkes and Brumley with my own very radioactive suggestion: that we work with *The New York Times* and give them an exclusive.

The suggestion was not well received. For both Brumley and Fawkes, their loyalty to the Administration made the *Times* an odious publication. Brumley, who commuted two hours each direction from his farm outside of Richmond, Virginia, once apologized that given his time constraints, he followed the news by listening to the "leftwing" National Public Radio. Both men counter-suggested that we should try and get a story in the *Wall Street Journal*. We all understood the *Washington Post* would be loyal to NASA and State, and not to lowly Commerce, so that was out of the question. The beat writer for NASA, Kathy Sawyer, would more than likely slant the story to side with the space agency.

In favor of my suggestion was that I knew Bill Broad, a young up-and-coming reporter who knew the Washington turf intimately. Today he has two Pulitzer Prizes and a reputation for handling influential science stories. He was one of the first American reporters to cover the creation of the Soviet organization Glavkosmos and its efforts to target American markets. In one piece just a few weeks earlier, Broad had quoted the head of Glavkosmos, Alexander I. Dunayev, as declaring that "our task is to commercialize the space program." That struck me as pretty powerful. We had declared the same goal numerous times, but never seemed to get off the dime. The Soviets were moving fast into territory that NASA refused to enter.

With hesitation Brumley agreed. He did this without telling the Secretary and without my being authorized to speak to the *Times*. Brumley's instructions were delivered in his best Marine cadence. I was to engage Broad before the license was granted but not reveal any details. If he agreed to work with us, I would give him an exclusive once the license was issued. If it were revealed that I had worked with *The New York Times* to break a story designed to embarrass others in the Administration, Brumley would consider me off the reservation and leave me to my own fate. It was a serious risk. Now, I too was way out on a limb. In retrospect, there were a lot of reasons to say no, but we were all caught up in the excitement of the project.

That same day I put in a call to Broad. I followed Brumley's instructions to the letter. The reporter listened and agreed that if the story had the national significance as promised, he would take it on. We promised to reconnect in the New Year.

One tiny detail remained: the final writing of the application. What to put in and what not to put in was scrutinized with biblical obsession. A nagging worry concerned an accidental discovery by NASA of the application. Maybe a coordinating clerk would see the word space, or space station, on the cover, and send it over to NASA. One night over beers that final problem was solved. We decided to de-emphasize the fact that the export license was requesting a research project to be conducted in outer space, aboard the Soviet space station. After copious revisions, the final application requested permission to conduct industrial research in a Soviet laboratory. That the laboratory was in space was not mentioned until later in the application.

While working as a free-lance writer I had learned that space holds a special place in our national psyche, and in the hearts of politicians. Without a doubt had the research laboratory needed by Payload Systems been located in some suburb of Moscow there would be far less fuss. It was, after all, now only a black box. No one should really care. But because it was destined for space it would be held to far closer scrutiny. There would be wounded pride expressed by politicians and worried anguish from NASA and the aerospace contractors. It's difficult to explain-but we react to space as if having never recovered from the shock of Sputnik. The government insists on controlling all aspects of space, but once they have their control, interest seems to be lost and nothing happens.

The year was drawing closed with a bang. We were all exhausted yet exhilarated. The strategy had come together; the legal determinations had been written; the bigwigs had quietly signed off. Now the request to be granted approval for working onboard the Soviet space station was in the hands of Hank's colleagues. There was still no leak. A dozen people knew of the license application and it would take only one phone call or a slip of the tongue at a Christmas party for NASA or State to learn of our trickery.

So we waited.

Chapter 4: The World Turns Upside Down

The hurricane warnings were up but I still didn't understand. Everything done so far was preparatory work. Once the news was public we would bear the full gale force winds of the NASA counterattack. It was something I had yet to experience, and didn't comprehend that for many in Washington, we were upsetting the correct order of the world. Three key bedrock beliefs were about to be rattled and turned upside down.

Firstly, the president sets foreign policy not junior level Commerce Department officials.

Next, innovative cutting-edge researchers do not work with, nor do they even have a need to work with, the Soviets.

Finally, NASA oversees all space programs, especially those involving international partners. No, correction: NASA oversees all space programs. Period.

For supporters of NASA like Congressman Bill Nelson the Payload Systems deal turned their world upside down. Photo: NASA

The real fun was soon to come.

In early January Arrott was on the phone from Boston. The company had received guarded news from Flowers that the Pentagon had reached a decision on the license. My office was keeping the news, whatever it was, good or bad, from me. But I caught Hank smiling, so I figured the news was good, but I couldn't tell Arrott that not only was I out of the loop, but my tea leaves had become the facial movements of that veteran poker player of trade policy Hank Mittman. Then I heard loud exchanges between Mittman and Fawkes. Could be bad. It was Flowers who finally gave me the news. She called me, all playful. "Got something you want to know," she teased in her high, laughing voice. The Pentagon's Export Bureau had signed off. We were halfway there.

Smiles or not, I was still pretty nervous until all the paperwork had been completed. Brumley had already taught me that lesson. One quiet afternoon Brumley recounted the story of how he and others in the Administration had managed to pull the space shuttle fleet out from launching military and commercial satellites after the Challenger accident. It required persuading Ronald Reagan to rule against NASA, something the President was never comfortable doing. The simple answer is that Reagan never, really, ruled against NASA on that issue. He followed staff instructions. The multi-agency National Space Council debating the issue was in the end being directed by Reagan's chief of staff Don Regan, the hard-talking former Wall Street executive. Finally, accepting the advice of then-Secretary of Commerce Malcolm Baldrige, over that of the vehement objections of NASA's James Fletcher, Regan walked into the Oval Office carrying the Presidential Directive removing NASA from the commercial launch market.

That piece of paper represented at least two years of bitterly fought work. The chief of staff explained the situation to the president. Re-launching a private American launch business. Getting military satellites onto more reliable unmanned launch vehicles, instead of the fragile manned space shuttle. Helping NASA focus on getting the shuttle back to flight status after the Challenger accident. The president interrupted his chief of staff with a wave of his hand. "I got it Don." Or words to that effect. With that Regan handed him the final page of the Directive, with two choices. Either to "Agree," or "Don't Agree." The President without hesitation signed with a flourish.

Regan took the paper back and gave it a quick glance. The President had listened to Regan but gone ahead and signed to allow NASA to continue launching satellites. Regan later told Brumley he wasn't sure what to do. Walk out of the Oval Office defeated? Instead, he decided to wing it. "Mr. President," he said, turning back into the Oval Office, "you've signed the wrong box." The president apologized. Crossed out the "Don't Agree" box, initialed the correction with a simple "R.R." and ticked the other box. With that NASA was barred from launching satellites. An entire new American industry flourished, the Pentagon would soon have a robust range of launch vehicles to use. And we stopped losing business to the French and their launch vehicle, the Ariane. Just like that. Over NASA's objections.

Bob gave me a copy of that Directive as a reminder about counting your policy chickens before they hatch. Clearly visible is the bold cross out and the "RR" correction. That's in the end how policy is made. You have to control the last guy who carries the document to be signed.

Mittman was now in the final handholding with our Commerce export officials. That was the reason for the tense closed-door meetings. Hank had warned me a few weeks ago to avoid going near the Export office or talking with them. There could be no overt sign that we had influenced their decision. The guys working the license had very delicate personal concerns that were being transmitted to Hank and from Hank upstairs to Brumley. Would the fifth floor, meaning the political appointees, stand by their decision? These veterans wanted assurances no one in the building would start second-guessing them. Would they be protected if scapegoats were needed? These questions were never asked, and hence no answers exist. But the dialogue was going on nonetheless. Soon we got the news that the Commerce Departments' Bureau of Export Control had approved as well.

The request by Payload Systems to export a black box to the Soviet Union's space station Mir had been approved by the Reagan Administration. It had happened fast and the news had been kept quiet. We, the newly formed Office of Space Commerce, had a major victory for industry, and Payload Systems had the first ever commercial deal with the Soviet space program ready to be implemented.

My warm and fuzzy feeling about how nice is the world of politics was about to smack hard against reality.

February would be a difficult month, even if all went well. I was scheduled to go in the middle of the month to a conference in Montreux, Switzerland. Also going was George Economy. Present at the event would be officials from the Glavkosmos organization. Finally I would have a chance to meet the Russians. First there was the need to get the word out on the license. My worst fear was that it would leak, and we would learn of a NASA news conference denouncing the program. Maybe they wouldn't care, but given our lack of resources, and that no national reporter knew of the Office of Space Commerce, it was essential that Payload Systems and Commerce frame the debate, not the big-boys.

Arrott and I scoped out the key objectives for the *Times* article. From his perspective, he wanted the pharmaceutical industry to understand that Payload had commercially negotiated for operational and routine access to space. On our side we wanted the Administration to realize we could implement with the Russians a economically efficient test-bed for the planned space station Freedom. We didn't view this as a trick, but just plain common sense. If industry discovered a medical breakthrough via the use of the Mir, everyone at Commerce, including Brumley, would support fully the best damn medical laboratory in orbit. But, if nothing developed, well then, let's be more realistic and honest as to why we were spending billions on a space station.

As agreed, without informing Fawkes, I went up to New York in late January and sat down with the *Times* reporter. We met in a bustling midtown coffee shop. I provided Broad with the details. It's funny when first meeting a reporter whose articles you have long been reading. Usually, they are younger than the image in your mind. Broad had already written much on the politics of the scientific world. But staring intently at me was an earnest young man whose proclivity for detail quickly became clear. Broad focused first on the Washington turf. He was trying to understand who won from this. That of course, was Commerce and Payload Systems. I tried to focus him also on the pharmaceutical market specifically and American industry in general, to look beyond the sensationalism and instead on the real need for industry to evaluate whether NASA's claims on micro-gravity were valid. His previous stories in *The New York Times* had focused on the big milestones, everything from the Soviets sending tourists into space to launching our satellites on Soviet rockets to supplying once-secret images from satellites.

With Broad, I wanted to focus on a very narrow view, that this project showed industry would pay for regular access to space, indicating that we should build a small space station for industrial use and not a Cold War multi-billion dollar station that would take years to realize. For Brumley, the loss of prestige by NASA may be enough of a victory, but I didn't want the story to cast too wide a net. I was now worried about Bill Verity. The last thing I wanted was anyone to link this to the Verity agenda, if there was a Verity agenda. Broad asked a few questions about Verity, but I answered with a nonchalance that was not at all genuine. The reporter also thought through who had lost. NASA and State Department of course. But Broad astutely also focused on some scores that Brumley was seeking to settle with others. He adroitly sketched in his mind a political roadmap of who would be angered by the news and who would embrace the development. It was a lesson in real-time journalism to sit over coffee, while a reporter as skilled as Bill Broad mulls out loud the different possible editorial slants for the story.

I left with Broad the contact numbers for everyone involved on the working level. In the days to come Broad called once. Then there was only silence. There was no clue when, or if, an article would appear. Both Fawkes and I were by now nervous wrecks that the news would leak and the chance to frame the news would be lost.

Fawkes was also making me an emotional basket case on whether I was going to Switzerland. Yes I was; no I wasn't. Yes I was; no I wasn't. Fawkes had reached the point where any major decision was just about impossible. He was working at least 12 hours a day, six days a week, sometimes longer. Mittman and Fawkes were fighting NASA on space station; the White House was issuing a new space policy that month that had been years in the making; and Bruce Kraselsky another member of our team, was working the PanAmSat and Intelsat issues.

On the day I was scheduled to leave for Switzerland, I packed my suitcase not yet having been given permission to go. I hung out all morning outside Gregg's office, waiting. His secretary interrupted an ongoing day-long meeting and asked if I was going. He couldn't answer her. Finally, it was time. We had fifteen minutes to call a taxi. I walked in and he asked one question: "Got your suitcase?" Fawkes needed a break. This man was stretched too thin.

I arrived in Switzerland, took the train from Geneva and soon saw the Alps. It was a glorious moment as it was my first time in Europe. That first day I met with George Economy and his girlfriend. It was the day before the conference, so our time was free. We had a great time wandering about the beautiful town with both palm trees and snow-capped mountains. Both enjoyed my discomfort experiencing jet lag for the first time. After dinner I finally crashed in the bed. Then the hotel phone was ringing. It was my girlfriend Dana McLeod - the Payload Systems' story had been published. On February 21st, 1988, the *Times* broke the news. Page one. Above the fold. Right hand side.

The article was straight-forward. "For the first time," reported Broad, an American company has contracted to have the Soviet Union carry Western commercial payloads into orbit. The agreement is likely to spur similar moves by other companies, Federal officials said."

Broad tackled the frustrations of industry, when he wrote how "The plan is an indication of the eagerness of private companies to perform commercial research in space despite the two-year grounding of the American space shuttles. It also marks one of the biggest successes to date in the Soviet drive to market the Russian space program, which was once shrouded in secrecy." Quoted in the article were George Economy, Anthony

Arrott, and a senior Hogan and Hartson attorney, not Ann Flowers. Broad addressed the issue of the export license pointing out that "The plan to join American industry and the Soviet Union in a space venture for the first time won the approval of the Departments of Commerce and Defense early this month."

Our news was the scandal de jour in Washington. Dana called in the middle of the Switzerland night to tell me that reporters were calling our house. The White House was considering a news conference. Poor Secretary Bill Verity had been hounded on a golf course by a CNN reporter wanting his reaction. Florida Congressman Bill Nelson, who represented the Kennedy Space Center district, branded the news as reckless. Nelson had himself taken some heat for flying on the last space shuttle before the Challenger accident, but it showed his commitment to NASA. The fifth-generation Floridian fired off a letter to Verity stating how "shocked" he was to learn of the granting of the license.

We had drilled straight into an open nerve. Newspapers all over the world ran with the story. There was even a *Los Angeles Times* cartoon showing first a Soviet rocket blasting off with the caption "U.S. firm joins Russian space program for experiments that could lead to new pain-killing drugs! The next frame shows a pained bureaucrat sitting at his desk, identified as "U.S. space program." The punch line: "They won't have to worry about their first customer!"

There I was, sitting in my hotel room in Switzerland figuring out how to dial international to speak with reporters. It was my first encounter with the power of the NASA brand with the national media. Newspapers everywhere were scrambling to run the correct slant on the puzzling news. Many of the phone conversations with the aerospace reporters were no different than those with the NASA public affairs officers. The tone of the questions was the indignation of a scorned lover. How could we facilitate working with the Soviets? How could we support a research program in space without NASA? Doesn't this jeopardize the NASA space station Freedom program? Won't the Soviets learn our secrets?

I was really disappointed by the questions. After the Challenger disaster in January of 1986 the mainstream press was critical of the cozy relationship that had developed between space reporters and NASA. The beat reporters believed their colleagues had grown too close to space agency officials. Sometime that year I attended a workshop devoted solely to the question of whether the media had grown too lax with NASA. In the afternoon something pretty dramatic took place. Several reporters stood up and apologized for buying into the NASA optimism on space shuttle, and missing the bigger story.

Answering the reporter's questions from Montreux, Switzerland it seemed nothing had changed in that cozy relationship. Whether the reporters were from Texas, Florida or California papers, many had anger in their voices. That somehow we were the bad guys. The negative drumbeat continued rapid-fire. *The Times* broke the story on a Tuesday. On Thursday Congressman Nelson held a press conference. ''It is not in the best interests of this nation to become dependent on foreign entities for our space research,'' barked the Congressman, who was having a hard time adjusting to a situation where the space shuttle was grounded yet American industry wanted to conduct research in space. Nelson sputtered how Commerce had bypassed the State Department's International Traffic in Arms Regulations (ITAR) process, which ''has reviewed such efforts in the past.'' He was right. The Congressman was totally right. We had sneaked it past them.

It was, for me, training under fire to witness the full force and fury of attacking the status quo. To be honest, one of the cardinal rules of Washington had been violated. I should have demanded from Broad a heads-up when the article was set to run. Then Ann Flowers could have briefed the principals the night before. Since we did nothing, everyone in Washington was caught unprepared. How much different it would have made in the long run, it's hard to say, but for several days all those who supported us were frustrated with how unprofessional we had been in victory. It was my fault. Fawkes and Brumley never asked about the article, wanting to avoid any of their fingerprints on the story. That was something handled poorly, but I never made that mistake again.

Weeks later I finally could ask Broad why we had ended up as the number one story in the paper. "A slow news day" was his laconic reply.

The meetings with the Soviet space officials fared no better. In my enthusiasm, a productive meeting among new allies was expected. We had secured the license, and now it would be time to discuss how to take our admittedly fluke policy victory and, as they say in Washington, codify it. By that I mean how to win over enough Congressmen, vote by vote, to the view that taking small steps in our space station program would be good for the taxpayer and our country. Let's find out if zero gravity research was as fabulous as NASA proclaimed. Let's do some more small projects on the Soviet Mir.

The Russians saw it differently. Very differently. They too read *The*

New York Times. They too noted it was on the front page. Above the fold. Right hand column.

The Russians at the conference in Switzerland and those in Moscow concluded that the granting of the license, plus the announcement via a page one story in the *Times* was proof-positive that the Reagan administration was fully committed to opening relations in the space industry. George Economy left me to handle this one on my own. Payload Systems had what it needed. Policy was now off the table for the company. George went off to iron out operational details with a couple of the Russians. On the last day of the conference I met with the Russian delegation for the third time. It was a late breakfast get-together at their hospitality booth, where the Glavkosmos organization was marketing Soviet space services. The brochures were poorly printed, on cheap paper. The English was grammatically incorrect, but the sales message was unbelievable: we are ready to provide on a commercial basis goods and services associated with the Soviet manned space program and the Mir space station.

With these men I once again carefully drew an honest picture of the political situation. How they should not interpret this as a message from the White House. How the political foundation for cooperation had not yet been established. More work needed to be done. The Russians did not believe me and were confused by my dour view. In Moscow the senior officials at Glavkosmos, and at the other space organizations, and apparently right into the Kremlin now regarded the Payload System's project as having great symbolical importance, one that might decide the future path for space cooperation between our two countries. Their analysts noted that the Reagan Administration had used a small commercial company to open the door, and thought this very clever. In the event of a program failure, it would not hurt the Administration politically. "Yes, now we have the signal to work together," was all the Glavkosmos delegation leader would say after I voiced my warnings for some fifteen minutes.

In Switzerland the Russian officials wanted to discuss next steps. How to get NASA business, and when they might work with the major players, such as Rockwell and Grumman. One triumphant official pulled from his pocket his already well-worn faxed copy of the *Times* article and read back to me one point upon which everyone in Russia was focused. *"In the wake of the Mir agreement, however, some Western experts have speculated that the Government's position is softening and that export licenses might be granted for American satellites to fly atop Russian rockets."* If this was true, it would be worth hundreds of millions to the struggling Russian space industry.

We were speaking right past each other. I dragged George into the meeting to vouch for my opinion, but they were all smiles, handshakes and congratulations. In vain I pointed out that "some Western experts" meant nothing. It was pure speculation. No chance our nascent private launch industry would permit Soviet launches. My warnings were overlooked. That's how the Switzerland meeting came to a close. I was dreading returning to Washington. The whole damn city was furious.

And the Russians? They were dancing.

The situation gnawed at me and later I wrote an article on the mistaken reaction of the Russians to the granting of the license. In hindsight I saw clearly (isn't hindsight great?) the repercussions of the Russians mistakenly believing the Reagan Administration had given them that green light. I wrote in *Space Policy* that to the Russians, without a doubt, it "signaled that a commercially structured Russian program, one supported by market conditions, would be met with political and commercial support from the Americans." This view by the Russians was wrong, and cooperation suffered from the misunderstandings.

Others did not agree with my take. One such response came from Jim Oberg, the influential former NASA engineer who for years studied the Soviet space program. In his book, *Star Crossed Orbits*, he psychoanalyzed my warnings on miscommunication, writing how "Manber's pessimism may be exaggerated, in large part because of the frustrations he faced in his position on the frontier of many of the cooperative efforts."

Hell. I was being critical because that first miscommunication led to the later miscommunications with NASA, which led to the miscommunications with Lockheed and Boeing, and finally those with the first Bush administration and the wasted battles with NASA administrator Dan Goldin. The *perceived* shift on policy by the Americans was one reason for the subsequent confusion from the Russians, which has added layers of challenges for today's U.S. policy makers grappling with a still confused Russian space community. It all started with that first mistake in miscommunication.

But, for the moment, everyone was pleased.

Payload Systems had their export license. Ann Flowers had a major success notched on her resume.

The Russians had their first American customer.

Bob Brumley had sent his huge "screw you" across the bow of NASA. Gregg Fawkes could demonstrate the value of the Office of Space Commerce; as the small office had stepped up to the plate for a U.S. company. Bill Verity had delivered one more market opener for trade with the Soviet Union.

And a low-cost and operational path to space had been opened without NASA.

Chapter 5: The Mir Really Exists

The bus had come to a sudden halt.

"We are here," Anatoly Merscheryakov of the Licensintorg organization announced proudly, but none of the three Americans believed him. The bus had stopped before a bleak metal gate, about eight feet high and fifteen feet across. There was no sign of life. Four people, three Americans and a Russian, were riding to the Soviet Mission Control, known as TsUP, in a bus that sat seventy people. Only in a society that cannot place a value on the price of labor, and hence of the hardware it produces, does the use of a huge bus for four people make even the least bit of sense.

It was December of 1989, and my one year stint at Commerce had finished. We had lost some major battles; including creating a commercially man-tended space station that would become operational after just two shuttle flights. NASA and Congress had beaten back that idea, since like the use of the Mir it represented to them a threat to space station Freedom.

We had won a great success with PanAmSat, which sought to end the

Ann Flowers and Anthony S. Arrott with Licenenstorg guide Anatoly Merscheryakov

Photo: Jeffrey Manber

international governmental monopoly on long-distance telephone calls held with the Intelsat organization. We fought not only the State Department, but also the telecommunications division at Commerce. The telecommunication lawyers at Commerce were completely loyal to Intelsat and refused to give the company approval to use the satellite for international calls and data. Multi-millionaire businessman Rene Anselmo was impatient and refused to wait for the policy dispute to be settled. He went ahead and arranged for the launch of the first PanAmSat satellite with no permission to use it once in orbit. Pretty gutsy. There was nobody like Rene Anselmo in the space business. The official motto of his company, as well as in his life was "truth and technology will triumph over bullshit and bureaucracy." The company logo was his dog Spot raising his hind-leg as if peeing on a lamp post.

When PanAmSat was weeks away from its first launch, I had lunch with Rene in Georgetown. I asked why he was launching PanAmSat-1 without government approval. The satellite could not be used until at least two nations agreed. Rene answered that government officials can out wait the private sector. If you want to be taken seriously, he advised, do something bold and the bureaucracy will rush to catch up. A friend of mine traveled to tropical French Guiana to witness the monopoly breaking launch. The French Ariane-4 rocket had recently suffered a number of failures so mission insurance was too prohibitive for Rene. He elected to launch with no insurance. If the rocket blew, he lost all his money, some $100 million. As the countdown neared the blastoff, Rene's son turned to my friend and dryly commented how "there goes the inheritance." His son was wrong. PanAmSat has enjoyed commercial success and changed our lives for the better. The competition lowered the cost of calling internationally and also led to media platforms like CNN. I have a hunch it changed the son's life for the better as well, as PanAmSat later become a public company worth billions.

And the Office of Space Commerce had won with Payload Systems. After all the fuss from the Washington political establishment, the anger from NASA, the recriminations threatened by Congress, it was clear that the license was legal and there was no god-given right for NASA to be involved with a American company working with the Soviet space program.

Now senior officials in the industry waited nervously. If the pharmaceutical crystals grown onboard the Soviet space station were inconsequential when compared to crystals grown in normal gravity, it could very well dampen Congressional enthusiasm for the NASA multi-billion dollar orbiting laboratory. It was surprising to me how much of the emotional debate about reliance on the Soviet Union and concerns over technology transfer were nothing but fig leafs for protecting the eight billion dollar space station Freedom project. Everyone knew it. Friends at Rockwell and Grumman took me aside complaining that their personal salaries were on the line, as well as their retirement packages.

The same duplicity had played out with PanAmSat. Congressional opposition was fierce for letting a private company handle cross-Atlantic data traffic. Publicly, supporters voiced concerns about breaking up the Intelsat system, since it had worked since the Kennedy era even if the costs were high. Privately, it turns out we were, well, there's no nice way to say this, but U.S. agencies had long monitored the Intelsat data traffic and their satellite network was riddled with intelligence listening devices. No agency wanted to give that up, and so all sorts of other reasons were invoked instead of accepting the inevitability of the free market to triumph. West Germany was the first nation to give PanAmSat permission to operate. Britain was second and we, under President George H. W. Bush, were an embarrassing and reluctant third.

In 1989, a few months before the Payload Systems launch, Anthony Arrott invited me to be part of the official observatory team as required by the granting of the export license. One team was to be stationed at the TsUP Mission Control in the Moscow region and one team was to travel to the Soviet republic of Kazakhstan and personally place the black box filled with the tiny crystals into the rocket. I said yes immediately. After all, I reasoned, you don't get many chances to go to Moscow, so I might as well take this one opportunity.

So, here we were, in late December, on a huge unheated bus. On the bus with me was Anthony S. Arrott, physicist and father of Anthony Arrott, and Ann Flowers. We were the official observatory team for TsUP. Anthony and the other researchers had already left for the launch site in Baikonur, Kazakhstan, about a thousand miles north of Afghanistan. The other Americans at Baikonur included researchers from MIT, the University of Wisconsin, and Harvard University, as well as company engineers including Julianne Zimmerman, Bob Reinshaw and Bruce Yost. Pharmaceutical companies from different countries supported these researchers. Payload's team was top-notch and held no prior opinions on the merits of space research; hence NASA's fear.

TsUP was the beating heart of the space program, the location where the cosmonauts, some of whom lived in space for months at a time, communicated back to the earth. It was the Soviet's equivalent of NASA's mission control at Johnson Space Center. Located in the town of Kaliningrad, it sits north of Moscow. A closed city until Gorbachev instituted his reforms; Kaliningrad was a company town, one that owed its employment and health to the fortunes of NPO Energia. How times change. Today it is a suburb of the fast-growing Moscow, and the streets are jammed with daily commuters. The town would later be renamed "Korolev," in honor of Energia's legendary first general designer, Sergei Korolev.

In 1989 there were over 30,000 people working directly for Energia out of a total population in Kaliningrad of 200,000. Yuri Semenov, the general director of NPO Energia, would later complain to me that not only did he have to run the space program, but worry about "the kindergartens, the sewers and the power lines of Kaliningrad." Moments before, by peering out the bus window, we could see factory after factory-huge structures surrounded by long rows of low apartment buildings. Occasionally, there were clusters of pre-revolutionary houses that were built-up wooden shacks that had somehow survived.

Two years later Boris Artemov, a senior official of international relations for NPO Energia told a sad tale while entering the once-secret town. His family had lived in one of the old wooden houses for over 100 years. The house was torn down in the 1970's to widen the street for the 1980 Olympics, the one boycotted by America. "You see, right there, where the guard booth stands, that," he remarked, "was the family house." The car cruised by the deserted booth.

"Every time I see that booth I think of the apple trees in our yard and of the cook-outs with the family." Artemov explained that he and his wife were given a one bedroom apartment in one of the nearby apartment buildings, which might best be called barracks. "It is a good apartment, he ended loyally. "But not as good as one's own family home."

Our host Anatoly Merscheryakov was in his mid-thirties and dressed well for a Soviet. Though his clothes were old-fashioned and the colors all one shade of brown. He had an engaging smile, which was unique among the Soviet officials we met. Anatoly was insistent that we had indeed arrived at Mission Control, so we obediently got off the bus and waited, first a minute then two and finally five long minutes in the cold of an early Moscow December morning. As we waited workers silently walked past, even though it was not yet six o'clock in the morning.

The wait was embarrassing. "They know we are coming, of course," Anatoly said a bit nervously. When no one appeared in another minute, our host went over to a little intercom, like that in the lobby of an old apartment building, and pressed the button. In a moment he was in a heated discussion with someone inside.

"It is all right," Anatoly finally announced, "the gate will be opened."

Indeed, in a few moments the gate did start to open. Slowly. It was being pulled open manually by an old man. He was dressed in a thick winter coat, an inexpensive fur hat and heavy construction boots. Anatoly politely nodded to the gatekeeper, who threw a quick nod in return, and led us through the gate and into a small courtyard. In the still-morning darkness our trusty guide carefully pointed out the patches of ice and cracks in the concrete. Forewarned, we gingerly made our way to a heavy wooden door, with Ann holding on for dear life to Tony's arm. This then was the entrance to the Soviet Mission Control Center. No signs, no tourist buses, no plate glass windows, no visitors' center.

In Houston, site of NASA's Mission Control, there is the long row of obligatory hotels on a wide street called NASA Road One, across from the Johnson Space Center. A visitor can choose from a Holiday Inn, Sheraton, or Big 8. These are sandwiched between the offices of the major contractors that do business with NASA. IBM, Lockheed, Rockwell, TRW, and Boeing, all surround the sprawling NASA complex of dozens of buildings. There is a new multimillion-dollar visitors' center, a NASA museum, and several rockets, including a flight-certified Saturn once capable of reaching the moon, and at least three major parking lots.

Here in Russia, in Moscow, in the heart of Kaliningrad, home of the world's most operational space program, there is a cheap intercom, a heavy metal gate and an old gatekeeper. It was more my image of a castle in central Europe than the mission control center for the earth's only orbiting space station.

First Anatoly, then the delegation walked through the large wooden door. What was inside made us stop and gasp. The entrance had lowered my expectations, but it was clear, standing in the visitors' foyer, that if the Soviet Union orthodoxy regards space exploration as a religion, this then was one of its houses of worship. We were standing in a richly designed, church-like hall, with white marble walls and twenty-five foot high ceilings. The contrast between the entrance and the interior could not have been greater. Here was my introduction to the traditional Russian view of industrial design. The outside "face," the one seen by strangers, must be stoic and unemotional. Inside can be warm and accommodating; this was true, I was to learn, whether for Moscow buildings or the people.

After gazing at the chandeliers and the twenty-foot high statue of Lenin, we removed our coats and hats, and proceeded up the open staircase to the main floor. I noticed a bouquet of fresh flowers at the foot of the Lenin statue. Anatoly saw my puzzlement. "Lenin is a genuine hero to many in the Soviet Union, especially the older women." For the first time, I walked up the red-carpeted steps at TsUP. Behind us was a stained glass mosaic of the three Russian pioneers in space. Konstantin Tsiolkovsky, the 19th century teacher who was the world's first space philosopher and Energia's first general director Sergei Korolev flanked by Yuri Gagarin.

Vsevolod Latyshev, the head of information department, or what we would call the director of public information, was waiting for us at the top of the steps. Vsevolod described the work in religious terms. "This mosaic represents our Holy Trinity." Here was the existence of something that I had been reading about for the past five years. That the Soviets had a far more spiritual approach to their exploration of space than Americans. Space exploration is for them an extension of their culture in many interesting ways. There is, for example, a custom that when a cosmonaut prepares for space, he plants a tree at Star City to mark the occasion. There is now a whole forest of trees planted by cosmonauts of all nations, some now over fifty years old.

Bob Reinshaw, the chief payload specialist for Payload Systems, told me that when he was at Star City preparing for their launch, the head groundskeeper showed him the tree planted by Deke Slayton, who flew on the Apollo-Soyuz mission in 1974. "Please tell Astronaut Slayton," said the groundskeeper, "that his tree is strong and growing well."

We had arrived a few minutes before the scheduled launch so our hosts took us on a tour of the mission control room for the Mir space station. The entrance of the control room is lined with the pictures of all the men and women that have flown with the Soviets. The wall, of course, is filled mostly with the faces of Russian men, and one woman. But here and there one sees a Vietnamese, Cuban, French and even that of the three Americans from the Apollo-Soyuz docking.

There is a balcony over the telemetry screen; below us are the mission personnel responsible for the space station and its two crewmembers. Against the front wall is an impressive flight screen that provides ongoing telemetry from the Mir; the flight pattern of the space station; countdown for the launch of the Progress, and a video screen for live coverage of the launch site. The scene was all the more remarkable given that the space shuttle program was grounded because of the Challenger accident. We had no astronauts in space, no manned vehicles, no zero gravity research, no countdowns.

It was now just thirty minutes to the launch. We were all pretty excited as the wait had been more than a year. It's not often, I was learning, that a resolution in Washington policy is followed by some real event. And rarely a rocket launch. Behind us were a few space engineers, though not as many as during a manned launch. I didn't understand it at the time, but it is a custom of the Russians to assemble during a launch, in the event something goes wrong. In a society lacking in quality phones, and routine communications, the most efficient solution was to bring everyone together. The event becomes a social occasion; news is swapped, a little gossip, and then afterwards everyone heads to the café for some coffee and black Russian bread topped with a slice of smoked salmon or meat.

With little fanfare the clock swept down the seconds to the launch of the Progress unmanned rocket. Lacking was the feeling of tension at the NASA control center. Now, it was just one minute before the huge cargo ship was to blastoff. We were told that the winter weather at the site was horrible. That news was again a reminder for us Americans of the Challenger disaster, which resulted from a launch during sub-freezing conditions. At precisely 6:30, in a heavy fog in Baikonur, the unmanned Progress lifted off with the Payload System's black box onboard.

We learned later that the fog was so heavy that Anthony and his team on site were unable to see the launch. Under those circumstances no American launch could have taken place. An unmanned launch was obviously quite routine for the Soviets, as they perform them every month in all conditions. We were able to watch the flight trajectory of the rocket until it had reached orbit. At that point Anatoly and the other Soviets were jubilant-in a Slavic sort of way. "There," pronounced the mission director, "you can see it is done. Now, let us eat breakfast."

At Mission Control that early morning were some of the key Russian space officials. Moments after the launch we were greeted by the head of Glavkosmos, Alexander Dunayev, of whom so much had already been written in the Western press. His arrival was unexpected, even the Russians seemed surprised. Anatoly whispered that this was a great honor being shown to us

Dunayev was of strong interest to me, as the Houston lawyer Art Dula had traveled to Moscow and signed a number of exclusive agreements with Glavkosmos and Dunayev. Dula is the true Marco Polo of American-Russian space cooperation, as he had managed to gain the rights to broker American satellites on the Soviet Proton rocket, to name just one potential goldmine. The problem with the Dula agreements was that there was no chance the State Department would allow these launches without concessions from the Russians. Had Dula come to the Commerce Department for help, as Arrott did, the Administration would have declined. Payload Systems was potentially doable, but the Dula agreements with Glavkosmos were not possible in the political climate of the time. In other words, Dula was too far in front of American policy to reap the practical benefits.

But here we were, part of the delegation of the first Americans to do business with the Russian space program. One would think Mr. Dunayev would have welcomed us warmly. Thanked us for our support. But one would be wrong. Dunayev, with his square, sharp face and portly body, looked more like a street-hardened truck driver than an international businessman, and his behavior matched his looks. The Glavkosmos leader lashed out at the American intransigence in working with the Soviet space program. "We don't need America," he spat out through Anatoly. "We will survive based on relations with other nations if you don't move quickly." His view was the polar opposite of the optimism expressed months before in Switzerland.

I didn't like the man, but Dunayev was right about our slow pace. The rest of the world was moving out on space commercialization. In June, ads had been placed in British newspapers announcing the Juno project, which would select one everyday person to take a ride to the space station Mir. Thousands of British answered the challenge. The Soviets had hired Saatchi and Saatchi, the international advertising firm, to manage the marketing of the Juno mission. Also announced was a project to send a Japanese reporter to the Mir, reportedly for $12 million dollars. So why was Dunayev so angry? He had plenty of good business and exposure. It was a phenomenon I've experienced again and again with the Russians. There is no one people the Russians love to denigrate more than the Americans. Yet, there is only one people the Russians feel are "good enough" to work with, especially in markets like space. And that one people are, yes, Americans.

As I got to know senior Russians very well, it was apparent they took the American attitude to the Russian role in space as a personal insult. Never mind that the rest of the world was congratulatory on each historic

Soviet accomplishment. Our poor manners grated. After the successful mission in April of 1961 of Yuri Gagarin, the Soviet hero went on a goodwill mission all across the globe. The cosmonaut was invited, and given a hero's welcome, in every country of the United Nations, save one. That one country was the United States. The Kennedy State Department wouldn't invite Gagarin.

Sitting in a Houston hotel room in 1993 during the difficult negotiations between the Russians and NASA that was deciding the Shuttle-Mir program, Victor Legostaev asked out of the blue why we had barred Gagarin from visiting the States. "Jealousy," was my simple answer. Legostaev agreed. "It's a pity; he was a nice man with a warm smile."

The non-response from the Americans towards the Soviets revolutionary embrace of commercial marketing for their space program weighed heavily on Russian industry leaders. Dunayev stormed out in a huff, allowing other officials to come forward. We were introduced to, among others, Nikolai Zelenschikov, who would later be the Energia manager responsible for the launch vehicles that carried NASA astronauts to and from the space station. He asked some questions. I wish I could remember what he said, but can't.

No one in the West knew that NPO Energia's senior management, including even the old timers now officially retired, was coming to the radical conclusion that it, the operational arm of the Soviet program, should handle the marketing of the Soviet space program, and not a government agency. It would be like Johnson Space Center, and part of Boeing, and sections of Cape Canaveral, splitting from NASA headquarters and launching their own commercial business. The new head of Energia, Yuri Semenov, just didn't believe the future was with yet another Soviet marketing organization. Something more radical was required. The old men who had controlled the production lines, and the launch center, and the mission control, and the research arm of the space program, came together against Glavkosmos, and in favor of casting their lot cooperating with the Americans. This was all being fought over as we gathered at TsUP, and perhaps that was one reason for Dunayev's fit of anger.

Our visit also took place during a very critical moment in the history of the Russian people. In Moscow the second meeting of the newly elected Congress was taking place, and in our hotels, and in the restaurants, people, everyday Russians, were watching on the black and white television screens the historic heated debates and outcries against the Soviet system and even against Mikhail Gorbachev. It reminded me of the Watergate hearings a long time before, when everyone stopped to watch the sensational testimony of Nixon's aides. Anatoly would suddenly flip from tour guide to businessman to communist party loyalist without warning, as if he too was battling in his own mind the same sort of debates taking place right now at the People's Deputy Congress.

On the return the next day for the docking of the Progress to the Mir, Anatoly was pointing out the sights the bus was passing when he remarked that two of the streets in Kaliningrad are named for Sacco and Vanzetti. "You know who they are, don't you?" he added, contemptuously. Ann and Tony were silent, though that may have been because they were asleep, so I ventured to suggest that "of course I know."

He was unconvinced. "Most Americans have no idea about these men." Put on the spot, I dredge up some fuzzy image of how to describe Sacco and Vanzetti. "They were anarchists with ties to the Soviet Union."

"Ties to the Soviet Union," he snorts. "What ties could they have possibly had?"

"You named some streets for them. That is a tie enough."

Anatoly didn't answer.

We were heading back to the hotel after the launch when our bus suddenly veered to the side of the busy road and uneasily slid to a stop. The driver, without a word, got up and pulled a toolbox from under the seat. Anatoly asked him what the matter was. "The brakes have stopped working on the bus," Anatoly explained without a tinge of concern. "It will be a few moments." So there we were stuck. On the side of the road in a bus provided by Mission Control. The brakes had failed and our driver was underneath pounding at one of the hydraulic valves. Around us cars and military trucks flew past in the snow and ice coming within inches of the bus, half of which was on the side of the road, and the back half sticking out. I wondered aloud if they had something like AAA in Moscow. Anatoly didn't understand. "How could you call?" he asked. "There are no phones." We waited in the suddenly quiet bus, with the cold now apparent. Equally apparent was the reliance on oneself-there were no phones or emergency services. If something goes wrong, one gets under the car and fixes it on the spot.

"It was your Senator Fulbright that caused the Cold War," Anatoly suddenly pronounced. "Men like him could not envision a world where we Soviets kept the peace as well." No one bothered to respond; here was another one of these strange moments. A politician like William Fulbright is not thought of often in casual American discussions on the post-World War II era. We might talk of Stalin and expect Russians to talk about Truman, but not Fulbright.

The bus driver finally reboarded with a huge grin. The bus, he loudly announced in Russian, was fixed. And off we went back to the hotel, having completed our official role as "observers" of the Payload Systems launch.

Phone Call to the Future

Two days later the cargo vehicle docked with the space station, and the Payload black box filled with the pharmaceutical crystals would now be on the space station for more than two months. The docking meant the mission was now truly underway, and the beaming Russians soon gave us a nice surprise. Located in the Mission Control building, off to one side from where the computers and consoles are located, there was a special room reserved for family members and guests so that all are comfortable while talking to the cosmonauts. After all, these husbands and fathers and sons live in space for months at a time; it was not the short hops of the U.S. space shuttle.

The room looked much like an American suburban recreation room, with nice wood paneling and big comfortable chairs, a television set and large speakers for hearing the voice of your beloved cosmonaut. While we were shown the room, someone picked up the microphone and handed it to Sasha Patera, one of the Payload engineers who had been at the launch site, a young blonde-haired woman whose family had emigrated from the Soviet Union. The cosmonauts were expecting the call, and suddenly, in this sacred room, where families talked to their space explorers, Sasha was cautiously chatting to the spacemen while their answers boomed from the speakers on both sides of the room.

The evening before was my first experience at a launch celebration, Russian style. Anthony Arrott and the researchers arrived in the midst of the reception, in dramatic fashion, having been stranded in Baikonur for a day by the awful weather. Cheers went up from us and loud "hurrahs" from the Russian guests, and it was a call for more drinking and more toasts. I remember my toast that evening, my first toast in Russia. Trying to be profound, I remarked how wonderful it was that the Progress launch carrying the American research project would be the last launch of the decade. After all, it was December 23rd, 1989. What a great symbolical end to a time that had seen our two nations coming closer together. Alas, my toast was soon proven incorrect. A day or so later a U.S. military satellite launched, so the final rocket launch of the decade was a military mission. Maybe that was a more fitting symbol, after all.

We drank far into the night so that the next morning I was moving somewhat slow and welcomed the chance to sit down in one of the comfortable chairs in the special room. I think if left alone I would have been asleep in a few moments. But then, without warning, Sasha handed the microphone to me.

I tried to protest, but she insisted. "What should I say? I begged her.

"You're a quick thinker, say what you want."

Taking the microphone in my hand I wet my lips and pushed aside my Russian hangover. First I introduced myself. I heard my voice through the speakers and then echo out into space and then reach the tiny Mir space station hurtling over Asia and leaving us at a speed of several thousand miles an hour. Overcoming both the fear and the hangover, I then wished "for good business between our two countries." After a moment of silence came the reply. One of the cosmonauts thanked me in perfect English, either Alexander Viktorenko, or Alexsandr Serebrov, I'm not sure whom. When speaking one could hear the slight echo of a long-distant call, only this time knowing it was going all the way up to space was special. At last, I remember thinking; I was able to make a long-distance call from Moscow.

But what a difficult moment. Sasha wouldn't take the microphone back just yet and our hosts, enjoying the moment, motioned for me to continue. Everyone in the room was smiling and expecting me to say something else. I knew it was supposed to be something poetic, or dramatic, or witty. Against the odds, we had done what we had set out to do, and something more had to be said. So once more I ignored my dry throat and pounding head and weakly wished both cosmonauts, who had been living on the Mir since September, "a smooth return to the good planet earth."

I was wishing the same for myself.

A small bit of history had taken place. An American company had paid for a launch from Gorbachev's Soviet space program. The launch and commercial cooperation competed with the meetings of the historic Russian

Congress and Gorbachev's warnings to the Baltic's not to push independence any further. And sad to say, it also competed with the invasion of Panama by the United States, an event that genuinely upset the Russians. On the day of the invasion we had a stream of officials upset at the American behavior. They felt that at the very moment the Soviet Union was winding down militarily, it was short-sighted for us in the West to engage in military actions.

For the Russians in the space industry, the Payload System's project was huge. Though they understood this was a small company, they also still hoped that perhaps bigger companies and then NASA would follow. Eventually, within a few years, they did. But there were no American companies doing business at the time but us, and I could see first-hand the awful mistakes of Gorbachev's policies. The new central marketing organizations that had taken power from the old factories were becoming irrelevant themselves as the government became more market-driven. IBM, Boeing and Rockwell do not share a marketing organization, for example, so why should all of the Russian space organizations use Glavkosmos?

It was the same in oil, in the financial communities, in every industry. Gorbachev was approaching the market place with the eagerness of a reformer, but the philosophy of a Soviet bureaucrat. The pity, the shame, was that "all the reformers left their old factories to work for these new Gorbachev organizations," a Russian bitterly explained to me after the launch. This paradox could be seen over and over in the next year. "Now the reformers will be put out of work while the factories, still controlled by the hardliners, are benefiting from working with America." This was my first clue as to what Yuri Semenov of Energia was up to. The reform-minded Russian space engineers and officials had to confront the fact that the factories, most of whom still supported the communists, were gaining month by month as the power of Gorbachev and the Kremlin ebbed away. While those that had taken the personal risk to work for the new marketing companies were now in danger of losing their jobs.

This terrible irony became the norm in the space industry. It meant that any continuation of the extraordinary push by the Soviet Union to treat outer space as a normal marketplace would be left in the hands of the hardliners. It made the events of the next few years even more remarkable.

Chapter 6: Aren't Two Routes Better Than One?

Much about the Payload Systems program has been forgotten, but it had a huge impact on the Soviet side, and some positive influence on a few key NASA observers. For me, speaking to the cosmonauts aboard the Russian space station was an epiphany. There really was an orbiting space station hurling through the void of space. It really had astronauts living onboard. There was a mission control room that looked a bit like NASA's, only the mission controllers were far more relaxed. These were not seven day missions like the space shuttle but a home whirling through space.

Nor was the space station still that tiny tin-can, as so many had dismissively labeled the Russian station when it was launched in February of 1986. A second module docked to the station in March of 1987. Known as Kvant-1, the unit weighed over 20,000 lbs and was crammed with equipment for astronomical research and material science experiments, the sort of research Payload wanted to perform. And, just a month before we arrived in Moscow, another module had been launched to the space station by the Russian workhorse rocket, known as the Proton. Called Kvant-2, it contained a docking port and a sophisticated life support system. Clearly, the Russians intended for Mir to grow, room by room. The program had both cargo ships and manned ships. The bigger unmanned ship was the

Mission Accomplished. A happy Payload Systems team at TsUP with their Energia and Star City colleagues Photo: Payload Systems

Proton, the smaller one, being used for ferrying cargo to the Mir, was called the Progress. The Soviets also had the Soyuz rocket that launched the Soyuz capsule, or, on its own, could launch smaller satellites. All these vehicles could take off in inclement weather and, knock on wood, had suffered only rare accidents.

I returned to Washington on Christmas Day 1989 filled with evangelical fervor. There is no embarrassment in saying this; it was as if Payload Systems had pressed the button of a time machine and pulled NASA's plans ten years forward. That phone call from TsUP to a crew living in space was at least a decade away for the American space program. And not because the Russians were more advanced than us-rather we continued to make the wrong decisions, investing our funds in grandiose projects that suited only the contractors. The question for me was how we could take advantage of this pathway to space through Russia. There was so much we could learn from looking to Russia-from how to launch rockets in bad weather to keeping astronauts calm during missions that lasted months, and equally vital, how best to tap commercial markets to buttress the government role.

I spent much of the next year speaking to those in Congress, in the White House, at NASA, even the Pentagon, as well as in the industry to follow-through on working with Russia. With the support of the Commerce Department, we announced the creation of a U.S.-Russian Business Roundtable for space activities.

Yes, these were the Soviets. But anyone in 1990 could read the newspapers. The society was changing, falling over itself to realize economic reform. It was not just the kids on the Moscow streets wanting to buy your jeans or sell you Soviet paraphernalia. It was the Soviet space officials who bombarded us with questions about marketing, public relations, cost of printing brochures, and the cost of sales commissions. How did NASA sell their goods? How did NASA arrive at their costs? What would the Payload Systems project have cost if done within the NASA program?

The pace of change was dizzying. Glavkosmos was signing deals for space tourism. Advertisements in space were not out of the question, nor was using their station to better understand what industry really needed. How to explain to these new believers in market capitalism that NASA did not sell their own goods? How to explain that NASA only bartered for goods and services using a model for international space cooperation now decades old? Or that it would be impossible to have done a contract with NASA for a research project like Payload Systems?

Meanwhile, back in the States, the desire of an innovative NASA-contractor like Payload Systems to work with the Soviet Union should have sent a strong warning message to congressional leaders responsible for oversight of the space agency.

Something was terribly wrong when not only did industry have to travel to Russia to implement their research efforts, but the export license had to be sneaked through the Administration. But the true meaning of the Payload license was ignored. No one at the space agency or in Congress moved to use the Mir as a first step to understanding long-duration space missions.

Instead, the U.S. aerospace industry ridiculed the deal with the Mir, while strong messages went out from NASA quietly warning other contractors against trying a stunt like Payload Systems. Typical of the reaction from NASA was a senior official at Johnson Space Center who warned me to back away. "Freedom will be built," he promised, "and you just don't want to be on the wrong team."

Anthony Arrott was as excited as me. He spoke of serious interest from pharmaceutical firms, both in North America and Asia. We thought that logic would prevail and this would be seen as a win-win for the American aerospace industry. Ironically, we were all hoping for great results from those little crystals growing on the space station-that would create a huge push for having our own space station based on the merits, and not as a boondoggle project. Right after returning from Moscow, I wrote a piece for the industry publication *Space News*, calling for Mir to be used as a testbed for Freedom. For Arrott and myself, as well as supporters like Bob Brumley and Gregg Fawkes, it was never Russia vs. NASA, but how using Russian assets could help us develop a more cost-efficient and commercially realistic program. This view was lost on the aerospace industry, which viewed competition, whether from entrepreneurs, or now from a commercial Russian space program, as dire threats to their profit margins. The pushback to the use of Russian assets only grew in intensity, and we were all warned not to mess with foreign policy or NASA's major programs.

The publication of the article in *Space News* at least had the affect of creating an informal network of those supporting cooperation with Russia. Some were new acquaintances, like Chris Faranetta, who worked at the Space Studies Institute in Princeton, New Jersey, which was the brainchild of legendary physicist Gerard O'Neill, the founder of the basic ideas on space colonization.

Faranetta, a 29 year old with little formal space science or Russian technology education, had none the less absorbed all he could about the Russian space program. Faranetta knew inside and out the Russian rockets, technical characteristics of the Mir, and was well-versed in future Soviet plans. He approached me in Washington with a novel situation: he had met the senior Energia officials at a conference in Spain, and they were interested in having him as their marketing agent in America. The problem was that Faranetta was unsure exactly what to do with his catch.

Some large companies also expressed interest in working with Russia, subject, of course, to a green light being given by NASA. In the middle of 1990 we held our first meeting of the Roundtable and more than fifty organizations showed up for the meeting.

I also stayed in touch with Art Dula, who couched the historical moment in language comfortable for American hard-liners. We have a unique chance, Dula preached, to lock the Soviets into capitalism. "And once the joy of capitalism is experienced," would intone this Texas-lawyer-turned-Soviet-trader, "it's impossible to contain." Still, as always, bubbling beneath the surface was the question of just why one had to travel to Moscow for innovative space projects. This was not about cooperation for the sake of political expediency-no, there was a scent of desperation on the part of both parties. By 1990 the Soviets were fearful of the economic and political changes sweeping through their society. On the other side, American space entrepreneurs were fearful of the rigidity that was suffocating our space program.

And those Payload Systems crystals that caused so much fuss? The results were disappointing. The pharmaceutical crystals grew for 56 days on the space station Mir and safely returned to the earth in late February. Once back in America, the sealed black box was taken to the Brookhaven National Laboratory in Long Island, New York, where the crystals underwent x-ray diffraction and other analysis.

A year later, a second research package was conducted again on the Russian space station. Most of the conclusions were kept proprietary. I found out later that Arrott had paid for the first flight using internal company funds. The second mission was paid for by industry. That was true entrepreneurism. The research results were published in *Science Magazine*, in 1992, and attracted strong reaction from the scientific and engineering media. Writing about the results in *Nature*, the author wrote that: "Despite the many millions of dollars spent launching science experiments into the near-weightlessness of space, the results so far do not wholly justify the cost, says a group of researchers in the U.S."

The *Nature* author explained that "Gregory Farber of Pennsylvania State University along with colleagues from the Fred Hutchinson Cancer Research Center in Seattle, California Institute of Technology in Pasadena and the company Payload Systems of Cambridge, Massachusetts, have carried out two long-term crystal growth experiments on board the Russian Mir space station. In their experiments, only 24 per cent of the crystals grown were better than the best grown on Earth. Results from numerous experiments on the U.S. shuttle are even worse they say: only 20 per cent showed an improvement. These results are important," realized the journalist, "because NASA is using experiments in crystal growth as a major justification for building the international space station Freedom. Farber says it is 'very hard to justify building the space station' on the strength of protein crystal growth."

NASA supporters simply ignored the research results. Those in Congress continued to use the promise of microgravity research as the justification for the tens of billions to be spent on our own space station. There continued for years later to be talk of a new era of space manufacture and research, but it became more low-key as most scientists lost their NASA funding in the overruns of Freedom. And given the few flight opportunities onboard the erratic shuttle schedule the non-aerospace companies gave up. Nor did any other contractor dare follow Payload Systems' example and seek to conduct industrial research in Russia. The time of optimism for Arrott and Payload Systems turned to frustration.

Our continued reliance on one single domestic path to space caused many space entrepreneurs to leave the industry. Russ Ramsland gave up trying to use NASA to grow the next generation super computer chips. Among other pursuits he opened a Tex-Mex restaurant in London, which I stumbled onto years later. Joe Allen, the former astronaut who worked with Max Faget on the small space station, pursued a very successful career far, far away from space. Anthony Arrott would leave Payload Systems to explore, among other projects, innovative Internet projects in Italy. George Economy returned to the financial world.

Byron Lichtenberg also left at some point, becoming a pilot for a commercial airline. Still involved in space projects, he took the pilot's job because, Byron told me, he loves to fly and there were just too few opportunities with NASA. Payload Systems continued on as a contractor devoid of any more troublesome schemes. So too the Office of Space Commerce. Sometime soon after the export license was approved, Gregg Fawkes decided to return to Ohio.

Congress would later apply extraordinary pressure on the Commerce Department, forcing it to be a line item in the budget, instead of hiding within the Office of the Secretary. Once under Congressional budgetary authority, the Office became a typical administration department, and would never again take such a bold stab into policy. It now goes by the name of the Office of Space Commercialization.

A fair number of Reagan officials that had supported our export license later chose to work with the Soviet Union as it transformed into a more market driven economy. This included Bob Brumley, who spent years setting up a free trade zone in Russia's Far East.

For those still determined to consider space as a normal place to do business, the idea of a second pathway was just too enticing. With the thawing of the Cold War, the use of Russian space services became far more acceptable until finally, it was in many ways the most robust path to space.

What did we learn from this first step with Russia? Payload Systems demonstrated that Russia could serve as a supplier of hardware on a commercial basis. Equally illustrative, it was also clear that Russians had no desire or capability to undertake their own market-driven projects, but were ready to support those of other customers. No Russian lotteries were being developed to send a winner into space; no Russian journalists were willing to pay to be first before the Japanese, and no consortium of Russian pharmaceutical companies were willing to fund microgravity research.

I became convinced that using Russian rockets and other hardware as a platform for Western ingenuity was the way to go. Maybe this was the long-sought elusive formula that would re-launch public excitement for space. Key to my interest was that the Russians were employing American free market principals. Whether out of desperation, as so many scoffed, or out of principal, the path through Russia would require a far more market-oriented approach by whomever needed Russian support. This was clear by 1990. But none of us would have dared predict that the wave of capitalism we saw sweeping over the Soviet Union would eventually transform NASA. It finally did, but this sea change took years. First NASA and their supporting contractors had to accept two profound propositions. Firstly, that multiple pathways for exploration are better than one. And sometimes the best capitalists are those most recently to the game.

Section Two: My Boss Semenov
1992-1998

This ship literally wreaks (sic) of both history and character. It's a "fixer upper" all right but one you would take a long trip with in a heartbeat. The central command post (cockpit) has keys that look like worn ivory. Leather shrouds serve where plastic would now be chosen. Its overall character brings forth the image of the "time machine" from H.G. Well's classic. Signatures and instruction placards written by the hands of over a decade of Cosmonauts who maintained and lived in this true marvel of human achievement.

Dave

E-Mail by Astronaut David Wolf sent from Mir October 31st, 1997

Chapter 7: Joining the Space Mafia

It was in a dark café in downtown Montreal that I came face to face with Yuri Pavlovich Semenov, the legendary head of the world's oldest space organization. Chris Faranetta from the Space Studies Institute had shown that his connections to the huge Russian space collective were strong, and a meeting was arranged. And to drink. Faranetta stressed that drinking was an important part of the introductory phase, but for me the original question had been just why I would want to meet with the shadowy leader of Energia, other than for curiosity. This was a man known for his fiery temper, outbursts of anger towards the West and his tenacity in defending Russian interests. What could we possibly have in common? Meeting Semenov seemed a step too far. Too dangerous.

"He wants to market into America," promised this young devotee of the Soviet program, and as proof he showed me a catalog that listed, on some sixty pages, the hardware and services available commercially from NPO Energia. It was vintage Soviet marketing. An expensive blue hardcover book with cheap paper and poorly written English, it was nonetheless a memorable document. Advertised was everything from crystal growth for research, renting cosmonaut time, use of research modules on the space station, and paying to fly to the Mir. The existence of this catalog was like Portuguese sea captains offering their maps and sea expertise to the British or the Spanish fleets. Energia had far more experience than NASA regarding how to live and work in outer space, and now this expertise was being shopped on a commercial basis. I began a long-distance discussion with Energia and the Space Studies Institute over the merits of setting up an American office for Energia.

The Montreal meeting in October of 1991 took place as the world began to change. Two months ago the coup attempt to bring down the Gorbachev government had failed, assuring that the market reforms sweeping over the Soviet Union would not only continue, but accelerate. Closer to home, space station Freedom was an organizational mess. Billions had been spent with little to show, and the cost estimates kept going up and up, while an exasperated Congress lost patience with NASA. It seemed the bad news on the budget would never end. The latest unpleasant surprise was the "Assured Crew Return Vehicle", also known informally as the station lifeboat. The lifeboat would be used to ferry astronauts earthward in an emergency. It was a vital station component that had been conveniently excluded in the original cost estimates. In 1989 Lockheed and Rockwell announced the cost to be at least a billion extra dollars. Congress did not take the news well and pressure was mounting to reduce the cost of the program. But how? Maybe one answer, admitted some, was to use Soviet hardware. It was for me a bittersweet change.

Sort of reality was seeping into the space industry two years after the Payload Systems uproar, nonetheless the effort to save Freedom still seemed the wrong objective for opening the door to the Soviets. Saving a billion bucks on the black hole that was NASA's space station would do little to solve the deep rooted problems afflicting the industry. My interest in helping promote the commercial Russian approach to space exploration was not to prop up NASA's mistakes but to force them on a new and competitive path.

The Administrator of NASA, Dick Truly, himself a former astronaut who had worked with the flight crew for the Apollo-Soyuz docking in 1975, visited the Soviet Union in October of 1990. While there, he met with NPO Energia and was probably the first U.S. government official to run straight into the Semenov "pitch." Semenov directly proposed using the Soyuz with Freedom. The Administrator came away impressed by Russian capabilities to help defray the mushrooming costs for Freedom.

The old adage about "politics makes strange bedfellow" was certainly true here. Truly understood that the Congressional political support underpinning Freedom was eroding. An eight billion dollar investment conceived during the Reagan administration was now estimated at no less than 30 billion dollars. From the perspective of a few realists at NASA, it was either figure out how to lower the out-of-control costs of Freedom, or risk termination.

Whatever the motivation, Truly's interest was the change that the Russians had been waiting for since Payload Systems, but just how to reap any benefits was unclear. There were multiple paths into the Russian space program. Mikhail Gorbachev had decreed in 1986 the formation of Glavkosmos. This organization had the so-called exclusive license to market Soviet space services in foreign countries on a commercial basis. Following the orders from the Kremlin, space manufacturing organizations located across the Soviet Union obediently joined ranks with Glavkosmos, which was soon labeled by the international media as the Soviet space agency.

But Yuri Semenov didn't believe these changes went far enough and digging in his heels, refused to join. His reaction was more than a turf issue. Semenov was thinking bigger. The Soviet patriot and his advisors believed Glavkosmos was too small a move to forestall the collapse of Russia's space manufacturing base. Glavkosmos was, he believed, just another branch of the doomed Gorbachev experiment. Semenov understood better than the disconnected Kremlin politicians how dire the situation. Unlike Gorbachev, he was willing to consider the most radical solutions.

His decision to open Energia to commercial business was in no way predictable. Yuri Semenov had been appointed the fourth General Designer of Energia in 1987, after a careful rise fueled, it was said, by his wife's political connections. Certainly, Semenov was technically skilled, political connections or not. Most recently, Semenov was the lead engineer on the Soviet space shuttle, known as Buran, which flew only once because of the Soviet budgetary collapse. During this time the up and coming engineer juggled the huge technical and political hurdles with an aplomb that made him worthy to take the helm.

After assuming control of Energia, the Soviet General Designer unexpectedly began transforming his government organization into behaving more like a commercial company. He would operate the "collective" like a CEO, taking control of both the technical side of Energia, as the General Designer, and the business side, as the general director. His economic reforms went far further than the Kremlin.

His first priority was securing the cash flow necessary to keep open the production lines for the Soyuz manned spacecraft and the Progress cargo ships. The short term solution was to begin charging to ferry visitors to the Mir. In so doing, Semenov's Energia broke with the tradition born with the space age, that space services were an extension of foreign diplomacy, not a commercial venture.

Though the commercial flights of the Japanese reporter Toyohiro Akiyama and British biochemist Helen Sharman garnered the attention of the press, equally radical was Energia's successful marketing overture to European space programs. European governments were offered the opportunity to conduct their own manned space program on a commercial basis. The move was embraced by the European community both for the political support it provided to a struggling Russia, and as a means to advance their own space ambitions without depending fully on NASA. Early examples included the March 17th, 1992 launch of Soyuz TM 14, with the "third seat" of the Soyuz occupied by Klaus-Dietrich Flade, with the ticket being paid by the newly united Germany, or the Soyuz TM 17, which launched on July 1st, 1993 with commercial passenger ESA cosmonaut Jean-Pierre Haignere.

Semenov's behavior was that of a classic disruptor. This term came into vogue during the Internet age, when companies came and went in the blink of an eye, and innovation disrupted the status quo with startling rapidity. Semenov's business model swept away, or disrupted, the NASA model used since the time of John F. Kennedy. The American model called for a huge central space agency, and the exploration program as a political tool in the Cold War. Semenov sought to replace the status quo of thirty years with a traditional competitive marketplace for space goods and services.

Energia's reforms were far more radical than anything ever implemented within not only the Soviet Union, but also in America. The new head of Energia envisioned a free market where his private sector cosmonauts would ferry everyday people aboard his Soviet rockets, where passengers could live aboard the brand new space station Mir, controlled by his company. Energia's space engineers would conduct industrial research for foreign companies and governments; even shoot advertisements for consumer products. In short, the transformation of the secret Soviet program into an American style open and competitive marketplace. Services for a fee. Marketing. Advertising. Tourism.

There was no shortage of Russian critics. Some based their doubts on patriotism, and the unseemly dependency on foreign customers. Others looked for the short-term fix and publicly floated the idea of selling the Mir space station for millions of urgently needed dollars. Some hardliners in the Duma looked to America for their skepticism. If capitalism was so wonderful, if competition so beneficial, if the free market so critical to introducing new products and lowering costs, asked these Russian skeptics, why hadn't the Americans introduced market principals to its own space program? Why had Ronald Reagan, of all people, and his handpicked successor George H.W. Bush, continued to promote an overarching huge government agency like NASA to control every aspect of design, regulation and operation of space hardware and services?

This romantic notion of Semenov towards America's space industry mirrored that of the early reformers' excitement with the Payload System's license. It was hard for Russians to understand just how protectionist was the American space program. Semenov believed once he was inside the "tent," he would find a commercial market like all other American industries.

Even in the dark café, Semenov cut an imposing figure; one which commandeered attention. He looked far older than his 57 years, a product of a life of hard work and hard drinking. The General Director seemed uncomfortable within himself, as if ready to jump out of his own skin. It gave a frantic energy to his meetings; this was not a man with whom you would discuss the weather. I would soon learn how true this was when traveling with him, as he rarely slept for any length of time. Questions would be asked no matter the hour. If lonely or troubled by an upcoming meeting, the hotel phone would ring and a meeting hastily arranged in the dead of night. At 6 feet 2 inches tall, and weighing well over 250 pounds, he resembled more an old-style Chicago politician than space official. Semenov spoke in a sneer, speaking out of the left side of his mouth. His negotiating and management arsenal included a dozen different ways to express contempt, from a dismissive wave of his hand, to mocking tones, to a disgusted look on his face to a high pitched scream when exasperated.

Translating for Semenov in Montreal was his trusted confident, Boris Artemov, a small elf-like man with watery blue eyes who spoke English with an Australian twang, the product of his father's embassy postings. Artemov was clearly fond of Faranetta, and was pushing Semenov to establish an American office for Energia. It was a management style that I would soon come to know only too well. Semenov had to be pushed and prodded into any new direction, even if he too agreed on the matter at hand. It was a tactic that forced his advisors to fully commit and be held accountable. Semenov had also grown fond of Faranetta, having met him in June of 1990. Faranetta, along with an engineer named Eric Laursen, was hoping that Energia would provide a free launch for Lunar Prospector, a SSI project to send an unmanned vehicle to the moon. The General Designer had bellowed an exuberant "*Pushkin!*" upon greeting Faranetta when we came into the café. Chris had the same curly black hair of the 19th century Soviet poet, and it became his nickname to be used only by Semenov, and only when the General Director was in a good mood.

Chris had a dilemma he thought I could solve. Semenov liked him and clearly wanted to market aggressively into America. But American aerospace contractors like Rockwell and McDonnell Douglas, as well as Truly's NASA, gave Chris scant attention. Chris was not known to the space industry and there was nervousness towards doing business with an unknown young American representing the Soviet organization. Though officials of Rockwell International had begun discussions with him on Soyuz ACRV, and McDonnell Douglas officials, according to Faranetta, had already informed Semenov they were not interested in working with Energia. Faranetta thought my heading an American office might solve the problem. In the time leading up to our meeting in Montreal we began a dialogue with Energia on how to market into the NASA space program. I suggested an industry advisory board, aerospace vendors as partners, and Washington as the right place for an office. Chris had wanted to locate in New York, but it was far too soon for a commercial foundation. We needed first to win over NASA, and the commercial business would follow. Besides, too many Russian-Americans were running around representing their relatives in the space business from offices in Brooklyn. Some distance would be good. We had to be a top-tier Washington effort that was respected, maybe not liked, but respected by the contractors. It was these discussions that led to the meeting during an international space conference in Montreal.

I remember asking about his short-term goals. Semenov growled that he was tired of being a government organization; that had got him and his programs nothing. "Now I want to be private, like Boeing or Rockwell," he shouted, his huge right hand closing in a fist and pounding the small café table. "I want my own customers, my own business, my own profit." Was he willing to take on a board of directors and publish an organization chart? Yes. He promised to behave as would the CEO of a billion dollar aerospace company. I joked he would have to learn golf like so many Western executives. "I don't play games," Mr. Manber," was the curt response. Noted. Jokes were not a good idea.

Before flying to Montreal I had heard nothing promising about Semenov. Art Dula had warned me that Semenov and Khrunichev's General Director Anatoli Kisselev would soon be pushed aside by Glavkosmos. Khrunichev was taking a different path of reform than Energia, one less radical. It would behave more like a company but remain a government organization. Art Dula was being told both organizations would soon be swallowed whole by Glavkosmos. On this side of the world NASA officials and reporters had pretty well stated that Semenov was a pushy bastard. Nonetheless, I respected the path he was trying to forge, no matter his uncertain political future or his abrasive personality. I had been given a copy of a September 11th interview in the Soviet newspaper *Izvestia* in which Semenov had laid out his challenges and objectives in converting his Soviet organization into a commercial company.

"Many countries want to collaborate with the USSR on Mir," Semenov bluntly told the *Izvestia* reporter, "but our potential partners are frightened by the stiff position of the USA, which doesn't want to have dangerous competition in the space market." Semenov confided that "now we are looking at the question of originating a joint-stock company in the foreign capital to utilize station Mir as an international scientific laboratory." Semenov admitted that the idea of a branch office in, I presume, Washington, was serious enough that he had sent a "letter to the President of the USSR." His comments were in an article headed, "*Is it True that the Soviet Union has Decided to Sell the Space Station 'Mir'.*" Sergey Krikalev, was one of the cosmonauts living on the space station when the possibility of selling the Mir broke into the mass media. He inquired to the mission controllers at TsUP in Kaliningrad, perhaps a bit nervously, if selling the station to the highest bidder included him as part of the package.

Listening to Semenov in Montreal I respected him more. The Soviet general designer didn't mince his words. He had a vision-a space vision for a new time and new era. He believed Glavkosmos was pushing business for the sake of cash flow, selling away Russian technology to nations like India with no thought to the long-term ramifications. The proposal for selling the space station was a panicked act; one proposed by frightened Kremlin leaders who saw no hope for tomorrow. Semenov wanted to sell Mir services, and work with the contractors, NASA, and the Europeans to develop a program to the Moon and beyond.

We spoke about the stupidity of governments, and he figured his was more stupid than ours. I realized that Yuri Pavlovich Semenov, a crusty and loyal Soviet general designer, who had carefully climbed up through the Soviet system for the last three decades, had in some ways been transformed into a middle of the road American businessman. His desires now were for less government, lower taxes, and a chance to compete fairly in the marketplace. In the *Izvestia* interview he grappled with the core issue confronting a commercial Energia. "Many problems could be solved more easily if there was determined the rights of property for the space station. Now it is not clear who really owns this orbital station-and as a result a number of profitable international agreements are concluded very slowly." Semenov then took Gorbachev's perestroika and blasted it into orbit. "Economic reforms should take place not only on the Earth."

Secretary of Commerce Baldrige had voiced the same words on pushing economic reforms into space some seven years before in the Reagan cabinet meetings. But the Secretary of Commerce was an interloper, a critic whose leverage against the major contractors and the space industry status quo was weak and temporal.

It was the *Izvestia* article that convinced me that NASA might finally be facing not just competition, but the possibility that one man, and one powerful organization, was capable of transforming the monopolistic U.S.

space program into something far more healthy, and, well, American in structure. I told Semenov that if his goal was sincerely to become a private company, I would sure like to help. His response was a vibrant and loud "ohh!" and standing up, we finished off one more round of vodka.

The plan to set up in Washington moved forward rapidly. Chris would leave Princeton and remain connected to Space Studies Institute as a board member. This was not an easy move, as the Institute was led by Dr. Gerard O'Neill, the visionary Princeton professor who had invented the concept of a magnetic launcher known as the mass driver. O'Neill had in the 1970's become a popular crusader for the human colonization of space. To channel the public's support for this futuristic concept, O'Neill had founded an organization whose mission was to turn his visions into reality.

Also working at SSI was Gregg Maryniak, who was Chris's boss. He wanted no part in the Russian scheme. That overture was left to Chris, who grabbed it with all the gusto he could summon. Maryniak later worked on the X-Prize effort, which was so important in launching suborbital commercial space flights.

A six month exclusive agreement with Energia was signed, giving us time to pull together a business plan. Envisioned was a fully professional office, reaching out to NASA, Congress and industry. The first roadblock was that the Russians expected us to initially work for nothing. I refused. "But the Germans work for us without pay," pleaded Artemov, referring to a small representative group in, I believe, Bremen, Germany. "Not me," was my answer. After weeks of back and forth it was agreed that if other conditions were met, they would provide a budget for several months at three days a week.

A trip to Moscow was necessary to have a final round of discussions and then sign the documents. Boris called the week before. "The boss will agree if all goes well." That seemed too amazing to be true. Amazing or not, there was one last step required in case the Russians were truly willing to hire me. I needed to have some cover from the political community. I may have pushed for cooperation with the Soviets, but it was a far bigger leap to consider working with them, and taking Russian funds. So before leaving for Moscow I visited Mark Albrecht at the National Space Council, a White House office that coordinated space policy for the Bush administration. Also present was Courtney Stadd, who handled commercial affairs, Gerald Musarra, who later worked with the Lockheed Russian venture that marketed the Russian Proton rockets, and George Abbey, who in a few years would serve as director of Johnson Space Center, and would come to spearhead NASA's dealings with the Russians. But that was all in the future, and the idea of working with the Soviets was unsettling-to them as well as to me.

I arrived at the Old Executive Office Building, alongside the White House, seeking advice and a letter "blessing" my working with Energia. There was some hesitation until little-spoken Abbey grumbled something like, "why not? Better they work with someone we know ." And so a letter was produced by Stadd, on National Space Council stationary, congratulating me on my new position with NPO Energia. The letter was carefully filed away, in case everything went politically wrong and I was hauled before some Congressional hearing demanding to know why I had worked with the Communists. Once the office was opened, we even registered with the Justice Department as foreign agents, just to be on the safe side.

There was one more stop. I visited the Pentagon's Office of Strategic Defense, better known as "Star Wars," which was the project begun by President Reagan to develop a science-fiction type satellite system that could destroy incoming missiles. From this hush-hush office Michael Griffin was supporting using Soviet space assets by American industry. Griffin listened to my situation. "You will either be a hero or the goat. Good luck." Later, when Griffin became the NASA Administrator, we both had a good laugh over his prophetic advice. After all, he was right. I did become the goat.

Thus prepared, a week later I arrived in Moscow.

As suggested by Energia, I took a taxi to the Cosmos hotel on the north end of the city, where my room took in a view of a sweeping titanium sculpture of a rocket soaring twenty stories up into space. The Cosmos high-rise had been built for the expected flood of tourists for the Moscow Olympics, so it was fairly modern. Boris Artemov came to meet me, and a rickety blue Energia van carried me further north to the space complex. After enduring a wait at the main gates for a few very cold minutes, we were led into the historic conference room of the General Director of Energia. The conference room was in the main building of the fenced off Energia facilities in the town of Kaliningrad, located just minutes north of Moscow. Without traffic it took only about 20 minutes from the Cosmos, but as we got closer the delays would begin. Just a decade before, the entire town of Kaliningrad was closed, with access provided only to Soviet space officials and their families. The town was just then awakening from the slumber of being off-limits to foreigners and Soviets alike, and construction was sprouting up on the main roads.

Semenov waved me into the large room. It was designed with soothing parquet floors with jarring green hued walls, though a comforting view of birch trees brushed up against the large windows. With a sweep of his arms he shouted from across the room. "How's Pushkin?"

I had visited the historic conference room on my first trip to Energia for Space Studies. It was here that the decision had been made in 1957 to launch Sputnik. In 1961, again at this very conference table it was agreed to send a human into outer space. On that occasion, confided Artemov, "there was much perspiration among those sitting at the conference table." I asked why. "No one could answer the simple question from our Kremlin leaders. If it was possible to send a man into space, why hadn't the Americans already done so?"

Semenov and his advisors had used the occasion of my first visit to ask all sorts of questions not just about the proposed office, and the situation with NASA, but also about our system of government and industry. Over the next two years there were plenty of formal and informal primers held on all sorts of issues we take for granted. Participating in the discussions were senior Energia officials, mid-level engineers, the Energia lobbyists working in the Russian Duma, the old former chief designers now formally retired from Energia, and even local politicians. All supported the commercial direction being taken by Semenov and were hungry to understand better the American government. Questions from them were often basic. Why the office of the president was so weak, how Congressional bills were passed, how NASA's budget was developed, how the NASA centers were related to one another, and so on. For a mid-1992 visit I brought several copies of the *Wall St. Journal*. A full day was set aside for my explaining about stocks, about the Dow Jones, about capitalization. "But where does the money from the private sector originate," asked the world's most knowledgeable expert on rocket engines? Some questions I couldn't answer, to my embarrassment, including how exactly the Dow Jones is configured.

Present for the meeting to decide the fate of Energia Ltd. was Boris Artemov, Yuri Semenov and a third man, named Victor Pavlovich Legostaev. It was he who had asked the question on the Dow Jones, and typical of the space engineer that he was, he would soon come to understand the Dow Jones and the financial markets as well as any American investor.

Legostaev seemed the closest confident of Semenov, a man with deep blue eyes and a full mane of gray wavy hair. Legostaev was older than

Semenov by a couple of years. He had been at Energia his entire career serving as one of the Soviet's experts on docking. In bits and pieces I came to know that he worked on Sputnik, and then served as a junior engineer for the Vostok 1 flight of Yuri Gagarin.

Legostaev had also been a principal engineer on the Soviet's first docking in space that took place on October 31st, 1967. It was the first time in history that two spacecraft had docked automatically, without a crew present. As usual with Legostaev, there was a great anecdote to share. "It was a mystery, our first docking," Legostaev confided one evening, in that deep, deep, husky voice filled with rich timbers like a fine brandy. "You see, when the docking time approached, we were counting down the minutes. Everyone was somber. Failure was not an option. There was nothing to do but wait. Everything was preprogrammed and when the capsules came into range, we started to receive the first data for that orbit. The data told us the two spacecraft had docked! It was a miracle. But how? We never knew exactly what had taken place between those two vehicles while they were out of our sight. But, as I was a young man, I was happy."

I would come to spend many, many days with Legostaev and he became the closest I have ever had to a mentor. Not in a business sense, but as a teacher in the art of understanding the absurdities of life. There was always richness to his stories, as rich as his deep voice.

On this day to decide the fate of the Washington office, I wasn't invited to sit at the mahogany conference table. Instead I was ushered through the door on the far side, through which Semenov had entered. Stepping inside, I was surprised to find ourselves in a small hideaway office. There was a small wooden desk by the window. On the desk were multiple rotary-dialed phones, some red, some black, one yellow, that provided direct connections to the launch site, to the Kremlin and to senior military officials, as Semenov proudly explained. On the right was a light colored bookcase and one more door, which I would later learn led to a private kitchen area.

We sat down, Semenov behind the tiny desk no wider than four feet long. Here, offered Semenov, had sat Sergei Korolev, the first leader of the Collective. I realized the meeting was going to go well, if they were showing me, a foreigner, the inner office so filled with their history, but I wasn't expecting what followed. After thirty minutes of probing questions on our objectives and chance of success, Semenov indicated he was satisfied. "If there is cooperation with your country, we will work with you for years. If no cooperation, we will not do business for long." A typically blunt but honest appraisal. Two sets of final documents were produced by the secretaries waiting in the outer reception area, each copy in both Russian and English, laying out the terms and conditions for working together. Semenov and I signed four times. I figured the event was over, and I could now go meet some of the program managers who wanted to discuss setting up meetings with Rockwell and McDonnell Douglas, but I was wrong. Semenov went over to the bookcase and took down a bottle of vodka. We toasted to the business between us, and as I reached over to shake Semenov's hand, the General Director came forward…and kissed me. Then Legostaev kissed me. Then Artemov. Each man kissed me twice, once on each cheek. I can recall right now my surprise at how rough is the facial skin of a man. And I had six times to give it thought. No matter. It apparently meant the deal we had just agreed to was sealed far stronger than any written contract.

The physicality of working with NPO Energia was soon accepted. It was like being one of the rookies on a championship sports team. A whole range of traditions and superstitions to learn. Hard work and harder drinking. But no one did space better than this group of men. No one.

When I recounted how the contract was cemented to Faranetta, he just shook his head in shock. "Good god," he burst out laughing, "you've become a "made member" like the guys in the Godfather. You're part of the space mafia," he added, some wistfulness in his voice.

Chapter 8: NASA Signs with Energia. Maybe.

The momentum for cooperation continued right into 1992. It was a time of optimism after two years of little progress. No garish blinking neon sign announced a new day had come, but from Congress, the White House and even NASA, there was the sense that the economic and political realities were converging to force a self-evaluation of the American space program, and the possible use of the Russian opportunities.

Our new company, known as Energia Ltd. signed for the initial funds from NPO Energia in December of 1991. The funds were received from a Coca-Cola commercial project being undertaken on the Mir space station. We chose for our office a location near Dulles Airport in Northern Virginia called the Center for Innovative Technology. The entire building was a Virginia economic development project run by a levelheaded industry official named Mike Miller who I had known since his time as an early employee of Orbital Sciences Corporation. Orbital is one of the few space ventures that has gone from the drawing board to a publicly traded company. I have a lot of respect for anyone who helped make Orbital a reality. Working with Mike was Sandi Christensen, from my old Office of Space Commerce, so it was friendly grounds. The funny shaped, upside down looking structure across from Dulles airport was perfect for our plans, in a very underhanded way. Emphasized to the Russians was that this building was not commercial-it was a project of the Virginia government. The point being that a regional government was supportive of the efforts to market Soviet hard-

Two veteran NASA officials Arnie Aldrich (left) and Sam Keller (right) courageously supported working with Russia and Energia.

Photos: NASA

ware to the American space industry.

In March Chris moved into one of those faceless suburban developments that dot the highway south of Dulles airport. Before he made the transition from Princeton, we paid a courtesy visit to Space Studies Institute founder Gerard O'Neill, who was gravely ill with cancer. O'Neill showed up with a disapproving Tasha, his wife, who stopped us from shaking his hand, given the weakness of her husband's immune system. She curtly warned that the meeting must be short. So we stood with a few feet separating us, while I brought O'Neill up to speed on the project. How we had received the funds from Energia and established the American office. This alone was unheard of-that in December of 1991 a Soviet organization had wired its own funds to Americans. How we had begun meeting with very nervous aerospace contractors like Rockwell and Boeing, and that discussions had begun even with NASA. I rapidly explained everything that had transpired in the last couple of months.

A fatigued O'Neill gladly received the news. "But be careful," he warned, "make sure the Russians don't take from us." He ended the meeting slightly more upbeat. "OK, let's see how far we can travel with the Russians. Good luck." That was our last meeting. In April the brilliant physicist succumbed to the Leukemia and passed away. I was glad we had a chance to report that this one project supported by his Institute had taken off.

Winter Thaw

Chris and I began introducing the aerospace contractors to Energia Ltd. who greeted fairly warmly the idea of an authorized American office for Energia. No doubt they were prodded by NASA's interest in our office as well. A consensus was emerging from within NASA Headquarters to engage the Russians regarding use of Soviet hardware for Freedom. The two chief pieces of hardware being targeted were the Soyuz capsule for the lifeboat, and the docking mechanism, known as APAS, first used for Apollo-Soyuz. All this interest was being directed by a NASA official named Sam Keller. Soon after we began introducing ourselves to the aerospace contractors I received the first phone call from Sam Keller, a 30 year veteran NASA official, whom I had met several times. Keller had worked to create a program for exchanging Soviet and American medical astronauts in space, but the idea had gone nowhere. Now there was a real shot at cooperation. NASA was serious about looking at the Soyuz, Keller explained in his raspy voice. He voiced his intent to make sure the interest in the Soviets was done right, and was done through NPO Energia. Over subsequent calls, some during the day and some late in the evenings, Keller made clear he understood the history of Energia, the position of Semenov, and believed that NASA would do well to win over Semenov from the first steps of cooperation.

It was an unusual situation. Keller was the number three official at NASA Headquarters but ignored by many operational managers. His portfolio seemed to be both arbitrator of internal disputes, and, more importantly for us, the manager of the handful of medical research agreements continued after the Apollo-Soyuz mission ended. There had been no eagerness by up and coming NASA managers to take on responsibility for working with the Communists, so by default Keller had become NASA's Soviet expert. The rapid changes from perestroika and NASA's own falling fortunes changed the desirability of that role overnight. Sam Keller, ignored by so many within NASA for many years, was now front and center for the plans to lower the cost of NASA's space station.

Keller decided that given Semenov's commercial slant, the best path was to issue a contract for studying the two pieces of hardware. He reached out to others within NASA for help in drafting such a novel document. Before Keller flew over to meet with Energia, he turned to a fellow veteran of Apollo-Soyuz, a steadfast senior engineer named Arnie Aldrich. Arnie, who probably knew more about Shuttle than anyone, was now in charge of Freedom. Aldrich agreed it was worth exploring whether the Russians were in a position to help.

Legostaev was surprised that these two veterans of Soyuz-Apollo, as the Russians referred to the program, would rely on Energia Ltd. More than likely we could not have survived the fierce NASA attacks that would start by year's end without this early acceptance of the office. But Keller and Aldrich did trust us. On their side, the NASA officials more than once expressed surprise that Semenov and Legostaev were allowing two foreigners to become involved in Energia business. I was surprised as well. For those who understood the history of Energia, this was openness as radical as the glasnost sweeping the rest of the Soviet Union.

Keller and Aldrich understood the political power of the space station contractors on the Hill, and working with the Soviets had to be seen as a net gain or else cooperation would be squashed before it even began. Whether coordinated or not, I don't know, but the contractor landscape became clear within a few months. Boeing was focusing their attention on the Soyuz as a possible lifeboat, as did Lockheed. Semenov had met with Boeing in October of 1991 and pushed the idea with them on using the Soyuz. Rockwell, after some hesitation, concentrated on the use of the Energia docking system. I think they would have preferred to build their own docking unit, but the estimated price tag just kept coming in too high.

Helping start the delicate discussions was that the first wave of aerospace officials was mostly Apollo-Soyuz veterans, and as such was known to Semenov and Legostaev. However, no step was taken on the American side by any executive without the approval of NASA. Faranetta was shocked. "I thought the title of vice-president meant something." But it didn't. In meetings in California and in Washington senior industry executives, all with very impressive titles, carefully sought to map out with us just how to approach working with Energia, while keeping Keller and later Aldrich abreast of every single step. Sometimes this meant keeping us waiting while a senior vice president left the room to make a quick call into NASA.

A nice surprise in the first months was that the Energia official who would interface with the Americans on the docking unit was Vladimir Syromiatnikov, who I had come to know well from Space Studies Institute conferences. Vladimir was the archetypal crazy scientist, with wild hair and clothes often a mess. His nickname to his American friends was "the big cheese," since the Russian word for cheese is "Syr." Syromiatnikov loved the time he had spent in Houston during Soyuz-Apollo and valued highly his friends at Rockwell and NASA's Johnson Space Center. There was tension between him and Victor Legostaev, but there was no question as to who should head the outreach to the Americans on the docking system. I think the tension came from the rebellious nature of Vladimir and his fondness for many things American. His dacha, for example was styled like an old American log cabin. His home in the country and his flat in Moscow, just a few steps from the Cosmos hotel, were always open to us. Not so Legostaev, who once invited me to his country home, but warned I had to turn off my cell phone to avoid my movements being tracked. Vladimir's defense of his openness was voiced in terms of his primary contribution to Energia: "Docking is by nature an act of cooperation." Also upsetting to Legostaev might have been Vladimir's love of good, old-fashioned dirty jokes. I must have heard dozens of off-color jokes on the "act of docking" as applied to more "earthly applications." Legostaev preferred to keep a serious demeanor in public.

Not that "the big cheese" was a pushover to foreigners. There was an early negotiating session with NASA in Houston where Syromiatnikov was the lead Russian. At the usual insistence of NASA at that later time, I was not allowed into the meeting. There was worry on my part that Syromiatnikov, in his eagerness to do business, would give away far too much. This was in 1993 and the Russians were very, very eager to become part of the American space infrastructure.

Syromiatnikov could feel my worry. I remember his eyes turned cold and his voice became strangely flat. "Don't worry about me, my friend," was the stern advice. "I endured the siege of Leningrad. We ate dog until there were no more dogs. Then we licked the glue on the wallpaper." His steel-blue eyes sparkled as he absorbed my shocked reaction. "So, relax, we will survive NASA also."

Keller's immediate objective was creating the climate that would provide cover for Arnie and his space station managers. During this time I worked mostly with Sam Keller, and rarely with Aldrich, to craft a contract that would satisfy the Russians, the space station office, State Department opponents and Congressional supporters. .

Arnie was-and is-the archetypal right stuff engineer which NASA was once so capable of attracting. He speaks only when necessary and has little time for idle conversation. A few years later he and his wife Eleanor-a powerhouse in her own right within the industry-invited me to Arnie's NASA retirement reception. Held in a huge hall in Maryland on a military base, the hundreds of invitees watched as picture after picture showed Arnie over a thirty year span. As a young man working on Apollo, later in his career on space shuttle, helping fix up the shuttle after the Challenger disaster- and finally as an Associate Administrator overseeing Freedom. Nothing changed with Arnie over those thirty years. Not the glasses, not the white shirt, not the firm look of determination. Only a little graying of the hair. Arnie lived and breathed space and he knew when someone else did as well. I was in his house once and Arnie took me down into the home office, the walls of which were filled with NASA and Russian mementos of his storied career. Together we looked over the Russian souvenirs. "Those guys know how to implement," said the carefully speaking Aldrich. "When they say it will be done, it is done." Praise indeed.

A NASA agreement was needed that would bypass the traditional means of cooperation, meaning government to government. Waiting for State Department would have been a non-starter. That would mean a commercial contract was required. There had never been a contract between NASA and a foreign entity. For that to take place, Keller believed a supportive Congressional hearing was required to provide some political cover. This proved difficult. The one-day hearing ultimately involved maneuvering by the White House, the State Department, two centers of NASA and the Russians. Keller was at the helm, beating back the negative concerns of the State Department and those within NASA fearful of opening a dialogue with the Soviets. Keller seemingly had Truly's ear, and as the late February date approached for the hearing we were fed a constant stream of information on what Truly would say and why. The Administrator would declare his open support for considering use of Soyuz, and the docking system. He would also state his hope that industry would pursue using the RD-170 rocket engine, which later took place when Pratt and Whitney used the engines for Atlas vehicles that carried Pentagon payloads into space.

Everyone understood that the State Department would not support Truly's comments, but Keller stressed how these views did not reflect either the White House or NASA's position. By mid-February Keller and I were speaking almost once a day. I would relay his comments to Boris Artemov and sometimes directly to Legostaev. Then Artemov would send back their questions for Keller. Persistently the Russians had questions about the motives of Senator Mikulski, who was holding the hearing. When Semenov finally met the pugnacious Maryland Senator, he took an instant liking to her fiery temperament. This was a woman he could understand.

To the Russians, Keller pounded away on the need for Semenov to "state his belief that we can work together in a commercial way." This was not to be government to government, it was a new path between old adversaries. Just as Semenov hoped, NASA was responding to his commercial overtures. A time of faint optimism for all.

From Peter the Great to Yuri Semenov

Semenov and his delegation arrived a few days before the hearing. It was my first experience in handling a Semenov "road show." Faranetta mapped out every meeting, every stop, even budgeting the obligatory time for the men to go shopping. I came to understand that shopping represented something deeper to the Russians than we Americans could understand. It was their tangible window into Western integration. Having a lawnmower, or a microwave or an Italian sports jacket meant far more than we realized.

One Saturday afternoon in 1997 I ran into Dan Goldin, the then-current NASA administrator in the men's department at Nordstrom's at Pentagon City. Both of us were shopping for shirts. ("Only America," wryly commented one Energia engineer, "would name a shopping center for the heart of the military industrial complex. Russian people, believe me," the engineer promised, "we would not shop at KGB City.") Goldin turned on me; giving me a tongue-lashing regarding Semenov's perceived stubbornness towards NASA. At Goldin's request, that same day I fired off a long fax detailing how I had accidentally run into the Administrator while shopping and itemized his long list of complaints.

Monday morning the phone rang. It was a senior Energia official. He had one question regarding my fax. "What store and which department did you meet with Administrator Goldin?"

Accompanying Semenov to America was Boris Artemov, and Victor Legostaev, along with two officials new to me, Valery Ryumin and Alexander Derechin. Ryumin was a former cosmonaut who had risen high into Semenov's inner circle. He was a huge bear of a man, who seemed pained by the need for cooperation with the Americans. Though once in the elevator at NASA Headquarters, Ryumin suddenly proclaimed "this cooperation is good business, but it's just too damn far." I never took to Ryumin, in part because he had developed his own connections in the States, and Ryumin seemed to regard our office and its control by Legostaev as an irritant.

Alexander Derechin was a different story. Derechin was not what the Russians would regard as "Russian," more than likely from the Caucasuses in the South. Whispered asides suggested he was also Jewish, which if true may explain why I found his sense of humor and love of language so delightful. Over time he would replace Legostaev as my primary contact within Energia, as he rose to the position of Vice President of International Affairs. A tall lean man now mostly bald, Derechin came to understand America as well as any of the senior management of Energia. This was not saying much, but it was true. He would come to comprehend how different America was to Russia, but was unable to handle, control or manage well any business proposition that was intangible. It had to be hardware to make him comfortable.

The man worked without pause, and my phone calls to him at nine, ten or eleven in his Moscow evening were accepted and welcomed, always ending with "I am waiting for your call Jeffrey, call me anytime." Nothing describes his work ethic better than the time Derechin kept me waiting for a meeting we had scheduled weeks in advance. I waited in his office first for an hour, then two, growing angrier by the minute. His secretary gave up with excuses. Finally he showed, apologizing for the wait. Only the next day did I learn the reason for the delay. Derechin had suffered a heart attack that evening, had been rushed to the hospital where he spent the night, but had made sure to conduct our meeting before taking two weeks off to rest. Why didn't he tell me? Why didn't he just cancel? That was the Derechin that I watched develop over the years.

The visit by Semenov would turn out to be the first of god knows how many times over the next seven years we hosted senior officials from Energia, the other Russian space organizations and the Russian space agency. Within a year Energia Ltd. established a travel department and all Russian space officials on travel for NASA would coordinate with us. Even those attending meetings in non-U.S. cities, whether Tokyo or Paris or Bombay often relied on our office for assuring all went smoothly. "Smoothly" usually meant they were paid on time. Russians insisted on being paid a per diem whenever they travel, by Russian law it was fifty-eight dollars a day. And so there we often were, scrambling to assure that a Russian delegation arriving anywhere in the world would find the funds waiting, either at the hotel or in a local bank account. More than anything else, handling their payments forged a bond of trust. There were several powerful Russian government officials who disliked foreigners like us knowing their business but still, we maintained their American bank accounts and helped pay the bills. We never divulged where they had been or how they spent their per diem.

It grew into such a large undertaking that we came under the scrutiny of investigators probing money laundering between Moscow and New York banks. The federal investigators were puzzled because we were handling huge sums of money in reverse. Money from NASA and other international organizations into Russia, not the other way around. We took care of the routine and the unexpected. There were several deaths and a few scandals. Early on we met with the police departments surrounding Kennedy Space Center and Johnson Space Center and more times than not, any problems were handled delicately. I understand errant NASA and aerospace contractors were treated with the same courtesy by Russian authorities.

A celebrating Russian space engineer who finished a night of festivities with a bottle of vodka on the beach at Kennedy Space Center was one of our biggest public relations nightmares. When found collapsed on the sand early the next morning, the Russian was fully nude and believed dead. The ambulance arrived and as the body was being lifted onto the stretcher, it suddenly rose, Lazarus-like. The awakened "corpse" ungratefully punched the shocked ambulance worker and tried to run off. That one could not be kept under wraps and made newspaper headlines in both Florida and Moscow.

This visit by Semenov for the Congressional hearing in February of 1992 was the first for Energia Ltd. Strange as it sounds, our good relations with Semenov was helped by a paperback I was reading just before the delegation visited. In a store near Dupont Circle in Washington I had picked up a second hand copy of Robert K. Massie's classic biography on Peter the Great, the Russian Czar. Renowned for his hard drinking, strange customs, and devotion to his people, Petre the First, as he is called in Russia. Massie described how during their travels through Europe the Czar and his delegation would often wreck the homes of the nervous hosts, using antique furniture for the fireplace and leaving behind no wine or food. This seemed to happen when the host was fearful or dismissive of the Russians. But if the host set no barriers, and threw open with enthusiasm his mansion, then the Czar's delegation would leave the place exactly as it was found. Having been warned that Semenov could be a difficult guest, I wondered if there was an historic similarity in bruised Russian feelings.

For the first few days the Russian delegation was the guest of Boeing. The aerospace company was eager to demonstrate to NASA their ability to work with Energia management. When it was our turn, Boeing executives warned me to expect a huge extra bill. "You wanted him, take him," said the Boeing guy with a sigh. There was the mini-bar bill for dozens of small bottles of vodka and gin, dozens of pay per movie charges, and hotel fees for smoking in non-smoking rooms and so on. All the extra costs despite Boeing having insisted that the company would pay solely for the hotel rooms. Of course the Russians refused to pay for the extra charges. I thought it somewhat cheap of Boeing. After all, at stake was the chance to earn hundreds of millions in contracts and the aerospace giant was fighting Semenov over the mini-bar bill. This was repeated over and over by the aerospace contractors, who never gave thought to the cost to Energia for hosting the visiting American delegations during their time in Moscow. We were not Boeing, our pockets were slightly more shallow, but if Massie's depiction of the sensitivities of Peter the Great still held true, there was a larger issue at stake.

Swallowing hard, we informed Semenov that there were no limits when he was a guest of Energia Ltd. The General Director didn't nod nor thank us. Six days later I nervously approached the front desk at the Hilton near Dulles Airport to take care of the bill. There was not a single extra charge; and there never was. From Peter the Great to Yuri Semenov little had changed in how powerful Russians view their Western trading partners, nor how the West viewed with anxiety and downright trepidation the Russians.

A Dazzling List of Projects

The February 21st all-day hearing went very well. It was surreal to see the delegation of Russian space officials sitting before a panel of U.S. Senators. Just a year earlier such interaction would have been impossible, so holding the hearing was progress indeed. Senator Mikulski was at her best, blasting the State Department's desire to drag its heels on cooperation. For the Senator, it was all about jobs. Cut space station Freedom and her Goddard Space Center in Maryland would be in peril. Semenov also worried day and night about saving jobs. After a lecture from a senior Boeing official on the need to reduce Energia's work force by at least 30% because of the dire economic situation in Russia, Semenov later in private shook his head in amazement, rubbing his entire face as he did when truly perplexed. "I don't understand your system," he confided. "If I fire five thousand men, they will drink too much, some will leave their wives and criminal acts in Kaliningrad will rise. Then I have to pay to clean up the mess. If I keep them at their desks, I don't always pay wages, but I give them a hot meal once a day. Now these men must get up early every morning, put on a jacket and tie, go to work, and order is maintained. Isn't that better long-term than your system?"

There were other moments with Semenov that taught me how there are many different levels of free markets, and that the American free market is not always the most caring for the needs of a given community. One that sticks in my mind was a private meeting between Semenov and a very senior Boeing official, during which the official carefully mentioned that Boeing was planning to leave Seattle. Semenov was startled. "What about the local community?" Semenov asked, perplexed at the thought of the severing of community roots. "I can't give a damn about Seattle," was the answer.

The *Washington Post* on February 22nd reported on the groundbreaking hearing. Under the heading *Russians Offer to Sell, Lease Spacecraft*, the article described how a "top Russian space official went to Capitol Hill yesterday with an intriguing offer: The chance for the United States to buy sophisticated Russian spacecraft at bargain-basement prices." Semenov obeyed and did as Keller instructed, inviting NASA to lease the Mir space station and to explore using Soyuz as a lifeboat for Freedom. Typically for the ever-impatient Semenov, he went even further than Keller or Mikulski planned. "(Semenov) outlined a dazzling list of projects the United States and the former Soviet Union might one day jointly undertake, including diverting radiation from the atmosphere with huge solar mirrors and patching up the Earth's punctured ozone layer. We could present a number of interesting projects for international cooperation," said Semenov, and "I would submit that we've talked about this far too long, and should move from words to deeds."

John Boright of the State Department tried to throw water on the momentum Keller was engineering by cautioning how any talk of Russian agreements must be connected with the larger issues of defense conversation, non-proliferation and, interestingly enough, "creation of a successful, market-based private sector."

The hearing was a success, even though we all cringed at the reporter's depiction of any deal with Energia as being at "bargain-basement prices." Wasn't it enough that we would save billions, prop up Russian efforts at economic reform, assist NASA, and put to rest the Cold War space race? Must we also entice ourselves that this was all some giant space based flea market with incredible bargains?

On the same trip the delegation met with Rockwell and a contract was signed regarding studying the Syromiatnikov docking system. We also went up to New York and met with financial firms interested in assisting a private NPO Energia with raising operational capital. Also on the agenda was a news conference at the Washington Press Club. Semenov showed sophistication in understanding U.S. concerns, stressing that working with Russia would save U.S. jobs in the aerospace industry by allowing space station Freedom to be built at Congress-approved budget levels.

The night before Semenov's departure we sat in the hotel reviewing the trip's accomplishments. Semenov had met with Sam Keller, with Boeing and Rockwell, with the financial community, and with an attorney named Dennis Burnett, regarding protecting Energia's intellectual property rights. I was concerned that Western partners could take much of Energia know-how. We launched what turned out to be a hugely difficult effort to protect Energia's technology.

The General Director's disappointment that his bold suggestions for joint programs beyond space station were being ignored was palpable. He was forced to the realization that America was focused only on a space station. Bigger steps would have to wait. Semenov made an announcement. "I will help your NASA build this space station," he gravely promised. "Now I understand. It is the soul of the NASA organization, and until completed, we cannot move on to more interesting projects. Therefore I will help you do this," and here Semenov drew a large circle over his head, indicating a station orbiting the earth, "so that we can, one day, do this," and now his arm shot outwards, depicting the promised dream of a U.S.-Russian expedition to the Moon and Mars. This may all seem pompous, but it was nonetheless a critical decision. The powerful pragmatic dreamer had come hoping for more, but took back to Moscow a correct understanding of what America was capable of accomplishing in the near term.

NASA Demands: Surprise Meat Inspections

After weeks of back and forth discussions the first contract between NASA and a Russian/Soviet organization was ready. One of the last issues to be solved was how to classify Energia in the title of the contact. Keller and I discussed this on the phone. After several tries we settled, and this is from memory, on something like "NPO Energia, a quasi-commercial organization." What did that mean? Really, nothing, but Sam and I wanted to convey that this was between the U.S. space agency and an entity that was distinct from the just created Russian government. Legally, this was not correct. But philosophically, of course it was.

Keller asked me to carry over the contract, allowing Energia to review the massive document before their own arrival. So in June of 1992 I again found myself in the lobby of the Cosmos hotel in Moscow waiting for the Energia van. It seemed each time I visited the rows of slot machines had grown more numerous, filling the cavernous lobby like weeds. Now three rows of the clanging machines enticed some of the middle-aged South Korean, French and German tourists to risk a few rubles while waiting for their tour buses. Off in another hall construction had begun on what apparently will be the "Cosmos Casino." Maybe models will wear spacesuits and hawk space-themed games. Something to look forward to on the next visit.

There was a great sense of anticipatory excitement at Energia. This was not to be a repeat of the political one-shot of Soyuz Apollo. No one on either side wanted that. No, this was business. A real chance for the two nations to work together as true partners. Today a space station, tomorrow perhaps to the moon. Politically, it was critical for Semenov to show some tangible results with the Americans. His plan to stake the Russian space program's survival on the European and American markets was supported, for now, by the dozens of influential General Directors of the Energia chain of contractors. How long they would wait before selling out to other nations was an unknown question. Once again I was led into the conference room of the General Designer. Walking over to Semenov I placed the massive contract on the table in front of him. It was hundreds of pages. Its objective was to explore having Energia develop a Soyuz to use as the rescue vehicle for NASA's planned space station Freedom. For me, what was most promising was how the contract was not a political undertaking ordered from above. It came from within the NASA space station office and their desire to reach out to the Russians. It had the support of the Administrator, of Associate Administrator Arnie Aldrich and was blessed by the powerful senator from Maryland.

But the excitement in the conference room turned to frustration. First, the contract was only in English. NASA had not taken the time or the courtesy to have any part of the document translated into Russian. So any significant analysis would have to wait. Several days later, the frustration only continued. The "fine print" was traditional government-speak. It called for NPO Energia to adhere to all federal government regulations. The Russian organization would have to agree to surprise meat inspections and certify certain health regulations concerning their factory bathrooms. Equally Kafkaesque, the contract language stipulated no trading by the signatories with countries such as Iraq, North Korea, Vietnam and Cuba and the recently collapsed Soviet Union. To accept U.S. government funds for this study contract meant Energia was legally barred from doing business with the evil empire.

Alexander Derechin would cock his head to the left, much like an owl when he found something bemusing. He did so in the conference room several times over the four days before NASA arrived. "This is most interesting to us," he announced with a smile. "We are barred from doing business with ourselves."

Next question. How could Energia launch Vietnamese or Cuban cosmonauts if they signed this agreement? And, most critical of all, how could Energia maintain their pride if the organization was subjected to a bewildering range of U.S. federal regulations? Surely an exception could be made.

It was our introduction, all of us, to the challenge of bringing NASA into this new era. Sadly, the NASA legal department had added their standard contract clauses. The other NASA departments had also insisted on their requirements. The result was a standard federal contract reflecting not one bit that this was an historic effort by a small group of far-thinking NASA officials to both assure greater capabilities for Freedom and bring peacefully two former enemies together. I don't think either Keller or Aldrich had read the entire document, taking for granted the fine print contained in NASA contracts.

It was my exposure to one of the cardinal rules of Alexander Derechin. Alexander forced me to read every line of every contract we negotiated. Soon I was one of the few Americans, never mind I was working for the Russian company Energia, that read every page of the NASA-Russian contracts. For years I would be shocked when NASA officials from the International Department, or from Johnson Space Center, were ignorant of what exactly had been agreed to. At these times I would hear again in my head Derechin's sarcastic advice when early on, I too had been ignorant of the specific substance of the first contract. "Please, Mr. Manber," he would sardonically suggest, "read what we have signed."

Maybe he didn't read every page, but Sam Keller knew better than any NASA official how to work with Yuri Semenov. He put on a masterful performance. The problem with the federal regulations was put aside to be solved later between Alexander Derechin's department and NASA legal department at Headquarters. The first problem Semenov raised with the NASA delegation was the size of the deal. With his most pissed expression, Semenov laid into Keller. "You promised us this deal was worth a million dollars," bellowed Semenov.

Unruffled, Keller slowly glanced down to the front page of his copy of the contract. "It's for $999,999, Yuri," he said in a tone expressing annoyance. "That's close enough. We're just a dollar short." Keller explained that NASA regulations became even more onerous for contracts for one million and over, so better to keep it under a million dollars.

Semenov stared angrily at Keller. Silence fell around the conference table. All this theatrics for a dollar? But I was new to the ways of Semenov. This was about trust. Keller, a tall man well over six feet, stood up. I thought he was leaving. What I remember is Keller reaching into his jacket, pulling out his wallet. Leaning over the table he handed Semenov a dollar bill. Others recall that the delegation arrived with a crisp dollar ready to be handed over to Semenov.

"This makes us even," retorted Keller. "Nine hundred and ninety-nine thousand, nine hundred and nine-nine dollars from NASA and one dollar from me."

To this day you can see Sam Keller's dollar in the Energia space museum. Amongst the Soyuz capsules and artifacts from Sputnik and Yuri Gagarin, the American dollar bill is neatly stored in a plastic case with the inscription, "first dollar from NASA." But it was more than that for Semenov; it was a testimonial to a NASA man who kept his word. On the final day of the trip the contract was signed between a NASA contracting officer with signatory authority, and Yuri Semenov. Semenov himself later called it historic. For him, it was vindication of his steering his massive organization towards the open commercial marketplace. For those of us in America seeking reform, the significant fact was how for the first time NASA agreed to pay for services from a foreign organization. With this contract NASA was relying on a foreign vendor, not for political gain, but to secure needed hardware at the best price from the best manufacturer.

Unfortunately, one obstacle remained, and that was NASA itself. The actions of a few high-ranking officials would do little to change the mindset of the space agency. It took long months of hard negotiating to clear up the requirements for surprise meat inspections and factory bathroom inspections. By the time the issues were solved, the two chief NASA supporters of the agreement had been pushed aside by an unexpected and unwelcome correction.

Chapter 9: Propping Up the Russian Space Agency

Within a year all the groundwork undertaken by Sam Keller and Arnie Aldrich was effectively undone. By that I don't mean the cooperation with the Russian Federation, which came into being on New Year's Day. Working with Russia became critical as funding for space station Freedom came perilously close to being voted down multiple times by a frustrated Congress. Undone by the end of 1992 was the idea of NASA supporting the reforms of the Russian space industry via Semenov, as well as transforming America's own space industry into one of greater competition. It was frustrating to watch all the energy being tapped to bring about the long desired Russian participation, with little thought given to using this historic opportunity to change how we conduct international space cooperation.

The mid-course correction came about because of two new players in the Bush Administration. In April of 1992 Dick Truly was gone. In his

Dan Goldin and Yuri Koptev in an elaborate ceremony at the Air and Space Museum in Washington, DC, which undid the work of Keller and Aldrich. Goldin refused to work directly with Energia and placed future U.S. requirements in the hands of the fragile Kremlin. Looking on is (standing from left to right) Brian Dailey, Dana Rohrabacher, Barbara Mikulski, George Brown and Bob Walker. Photo: Jeffrey Manber

place Dan Goldin took over the helm at NASA. Goldin was little known to the industry, having spent much of his career as an executive for TRW's classified programs. I liked Goldin. The 51 year old was Brooklyn born, the son of Orthodox Jewish parents and bore a faint resemblance to actor Dustin Hoffman. He was also very brash. It was soon common for contractors to use terms like "strong-willed, confrontational, and decisive" in describing the new Administrator. His enemies, of which there were many, went further, referring to Goldin as "Captain Crazy."

Vice-President Dan Quayle, in his capacity as chair of the National Space Council, brought in Goldin despite his being a Democrat. Mark Albrecht and the Vice-President wanted Goldin to shake up the Freedom project, and not be intimidated by the entrenched aerospace contractors. They need not have worried. He took to the assignment with an enthusiasm rare in Washington circles. A PricewaterhouseCoopers report referred to "his impatient, demanding, intimidating management." Though agreeing with his desire to shake up the industry, I found the administrator mercurial in our infrequent discussions. Charming at times, arrogant at other moments, Goldin could not accept that perhaps his approach to Russia was wrong, nor even understand why others, such as myself, would even associate with Semenov.

Few of the aerospace contractors were pleased by the changes that swept through the industry. Maybe none. One went so far as to sue NASA. Goldin was deposed under oath, and particularly chilling is this excerpted exchange between Dan Goldin and the lawyers for Northrop Grumman:

Dan Goldin: In '92 (Space station Freedom) looked like a complete and absolute disaster. So I took it upon myself to think about redesign approaches and I talked to a lot of people, I don't know who, but a lot of people openly, inside NASA, outside NASA about that possibility. And then we came up with an approach which would be completely different than the Space Station Freedom approach. And within about a week of the time that that design came forward, I was called in to the White House and Congress and told to stop working on it because the contractors didn't want it to happen.

Q. In February of '93?

A. No. In the summer of '92, the Space Station contractors went to the Congress and complained to the Congress that the NASA administrator who was in charge of America's Space Program didn't have the right to look at any other design but Space Station Freedom.

Q. Who called you in from Congress, or who in Congress?

A. I didn't get called in. Let me say I got messages sent to me indirectly.

Q. Do you know to whom they went in Congress, whether it be NASA or contractor people?

A. All over Congress I presume. Contractors went there, interest groups went there. Sometimes they would even tell me they were doing it.

Q. You were earlier in your testimony today reciting the litany of travails you went through at the outset of your tenure as administrator, you indicated you had been accused of criminal activity. What type of criminal action?

A. I don't remember, but they did it.

Q. Who did it?

A. I don't remember who. It is just things (that) stick in your memory. It got pretty vicious here. People threatened me.

Q. Physically?

A. No. With reputation. They threatened me they would get me removed from office.

Q. This was contractors, NASA people?

A. Yes, yes, keep going.

Q. Congressman?

A. Yes.

Q. Which Congressman?

A. I don't remember which. In fact it came to me indirectly. No one openly said it to my face.

Q. Staffers?

A. Little birdies came to see me.

When the legal deposition became available, Derechin's staff carefully read the transcript. It was a confirmation that beneath the surface American politics could be as hardball as that in Russia. Energia management was also shocked that this sort of dispute could become public. For Semenov especially, hardest to accept was Western style transparency. Begrudgingly he adapted to western accounting and organizational procedures. By mid-1994 Energia financial data was being reported to shareholders. Maybe not with American transparency, but equal to that of a large French or Japanese aerospace company. Not that anyone at NASA ever read the Energia financial data. If they had, then the delay in producing key hardware promised to NASA by the Yeltsin government would have never been a surprise.

It's difficult to pin down any one description of Dan Goldin. Here was a government official who showed courage in seeking to shake up the industry. In his heart he was a romantic, who gallantly saw his agency as a symbol for bringing closer the governments of Russia and the United States. In this new world order there was no place on the main stage for a maverick like Yuri Semenov. Semenov's quest to elevate himself as a counterpart to the NASA administrator was viewed by Goldin as nothing less than the 'barbarians at the gate", a new sort of mafia attacking the integrity of the central government of Russia.

Goldin was supported in his views on Russia by the new head of the National Space Council. Brian Dailey, at 40 years of age an intensely emotional and ambitious Senate staffer, took charge of the Council just as we were reaching the agreement between NASA and Energia. Dailey, like Goldin, arrived with an entrenched view on the Russian political situation. Dailey would make much of his graduate expertise in Soviet foreign policy, and his close contacts with the defense community. The two men believed they understood the radical changes taking place within Russia and were impatient with those who disagreed.

Within weeks of Brian Dailey replacing Mark Albrecht, Keller could sense he was losing influence. Keller concluded that Dailey and his staffers were "gunning for me." He was right. Dailey later explained in a NASA Oral History project that as the National Space Council prepared to move forward with cooperation with Russia "we ran into the NASA bureaucracy in its fullest extent. There was a gentleman by the name of Sam Keller, I think his name was. He was strongly resisting this initiative and it was going to take forever. This was really a problem."

The idea of Keller being part of the NASA bureaucracy was ridiculous, but it only got worse. I never told Keller, but soon Gerald Musarra at the Council was seething in frustration over Keller's insistence on dealing with NPO Energia and promising to have him removed. I literally begged

Musarra to do nothing for as long as possible. Musarra relied on me at times for unbiased Russian advice and I relied on him for understanding Administration and NASA thinking on Russia. I knew as a result of these conversations that Goldin's NASA, with the support of Brian Dailey, had made the decision to turn away from NPO Energia and instead support a new organization, the Russian Space Agency.

Formed in February, just four months before, the Russian Space Agency consisted of no more than a dozen officials. It was lead by Yuri Koptev, a 53-year old bureaucrat from one of the Soviet's largest military organizations, from which he had overseen the Buran shuttle program.

Both Goldin and Dailey defended their decision to work only with Koptev by vehemently arguing that space exploration is inherently a governmental undertaking, and hence all agreements had to be between government agencies. They refused to accept the unique stature of Semenov. The dispute at times was emotional. There was one reception at the Russian embassy in Washington where Dan Goldin castigated me for supporting Semenov. In vain I tried to explain that Semenov was not receiving the promised funding from the Russian Space Agency. Instead, Semenov was transferring his commercially earned profits for use in the federal programs agreed to between NASA and RSA. This, to me, was an absurd way of doing business. Better to contract directly with the Russian private sector. The Administrator just refused to listen. "How can you carry this man's negative message into the halls of Congress?" Goldin demanded angrily, referring to our campaign to brief Congress on the value of the privatization of NPO Energia. Half an hour later I was cornered by Yuri Koptev, who in spitting Russian, stabbed his fingers into my chest, disturbed that I was speaking to Goldin on the impotency of his agency. Long moments passed until Semenov came to my rescue and the two men stood nose to nose, yelling. I beat a hasty retreat.

Sam Keller and Arnie Aldrich didn't choose to build a working relationship with NPO Energia because they liked Yuri Semenov, or because he was the prime contractor of the hardware that most interested NASA. They could have chosen to work with Glavkosmos, or they could have signed with the Russian Space Agency. Neither of these options made sense, not just for NASA's specific needs but also for the broader American objectives.

Sometimes a picture really does prove the point, and in this case we had several in our new office that provided unmistakable evidence of the importance of Semenov within the Russian space industry. In mid-1993 we left the Center of Innovative Technology and bought our own townhouse in Old Town Alexandria, just minutes outside of Washington DC. It was a major step for Semenov, reflecting the optimism that cooperation would be a long-term reality. The new office had three small floors; the top being the conference room. We hung two objects on the wall, the first an 18th century map of Russia, the other a photo of Sergei Korolev, the first general designer of Energia. The map was intended to reinforce the notion to our visitors that throughout Russian history the borders of the country had changed. Yes, this was a time of uncertainty with the collapse of the Soviet Union, but Russia would recover. The picture of Sergei Korolev was a reminder of the historic role of Energia within the Russian space program. But it was the Russian reaction to the plaque we glued beneath the Korolev photograph that spoke volumes.

In 1994 the prime minister of Russia arrived for negotiations with Vice President Gore. Victor Chernomyrdin was a crafty old oilman, one of the few of the old guard who nimbly made the transition to wealth when the Soviet Union collapsed. Space and oil were the two main topics for the high level discussions and when the prime minister's plane touched down at Andrews Air Force base it disgorged the leading Russian officials from both industries for the Washington discussions.

As Yuri Semenov came down the stairs of the plane in a mad dash, he slipped and hurt his leg. The next day, still in pain, NASA officials took him to the infirmary where the ankle was bandaged and a wooden cane was provided for walking support. That should have been the end of the story, but when he was boarding the plane to return to Moscow, Valery Ryumin handed me the cane and said, ceremonially, "this is the cane of your general director." Now, what the hell was I supposed to do?

Rather than returning it to the infirmary, I took the cane and mounted it on the wall near the image of Sergey Korolev. Then I had printed up a one-line inscription. Soon Russian governmental officials in Washington, even those on other business, would come to the office just to see the wall that had the cane from NASA and the corresponding inscription. The inscription, in Russian, said, "This cane is the only time that Yuri Pavlovich needed NASA's support." Embassy officials came and photographed the wall. Russian negotiators in town for State Department discussions came and stayed for the refreshments. My point is this: Semenov and his supporting dozens of aerospace manufacturing plants and hundreds of governmental organizations, should have been the catch NASA-and America-desired.

A final point on the cane. I forgot to tell Semenov or Legostaev or

Derechin. Some months later Semenov came into our conference room accompanied by a delegation of Russian government officials. With a puzzled frown Semenov went up and stared at the wall. "What's this?" He asked quizzically. "Manber, did you do this?" I looked over at the Russian government officials, who said nothing. Blank looks on their faces even though several had seen the inscription on previous visits and had laughed and snapped pictures. Now I was on my own in front of a short-fused Semenov. "Yes," I confessed, and told the full story. Semenov broke into a loud guffaw, joined immediately by the others. Good lord, never were these powerful government officials more like "yes men," than that moment. A little support would have been appreciated; and Semenov's power never clearer.

Goldin and Semenov Duke it Out

The formal signing of the NASA contract was a day of mixed feelings. Dan Goldin pounded home the point about not working with Energia by signing the deal with Yuri Koptev right on the first floor of the Air and Space Museum. Sam Keller and I stood off to the side, where an area had been cordoned off for the ceremony. Sitting at the signing table was NASA's Dan Goldin, the National Space Council's Brian Dailey, Senator Barbara Mikulski, other U.S. politicians and Yuri Koptev. Sam and I shared a mood that was both somber and jubilant. Our contract was being formally ratified by the United States and the Russian Federation, but what a mess waited for NASA. We both knew that. With the Bush administration having bestowed upon Koptev their blessing as the prime voice for Russia on international space matters, there could be no assurances that bilateral agreements could be honored. "Hey Sam," I finally joked, "how come no one at the table had anything to do with making this agreement a reality?"

Keller, who towered over me, just looked down. "Son, welcome to the ways of Washington."

A month later we could only hope for the best when on July 9th, 1992 a bitterly divided American delegation took off first for Germany and then Moscow on a trip designed "to gain a firsthand understanding of Russia's space program," as a NASA press release quoted Goldin. In reality, there was a grim determination to change course in a program now of interest to George H.W. Bush and Boris Yeltsin. Brian Dailey, Dan Goldin and representatives of the intelligence community were of equal mind to prop up

Boris Yeltsin's Yuri Koptev, no matter the consequences. Gerald Musarra kept me informed of the changes in thinking going on within NASA. With his permission I warned Energia that the American view was now to buttress Koptev. Semenov was in disbelief. How could the United States rely on an organization that lacked any influence? Semenov had just signed a multi flight commercial contract with the European Space Agency to pave the way for four more ESA astronauts to Mir. The funds would support the production lines for a solid year. By contracting directly with Energia, ESA's funds would bypass the bureaucracy of the feeble Kremlin government.

We still held some hope that the ties with NASA could be maintained, as Sam Keller and Arnie Aldrich were also part of the delegation. But despite the disagreement, I don't think anyone could have predicted the farce that followed. Disaster struck immediately. The first meeting after the group touched down in Moscow was with Yuri Semenov at the Energia plant. Semenov, in his blunt fashion, informed Dan Goldin that the days of governmental cooperation were no longer possible. The Russian government had no funds and hence there was no responsible government agency that could fulfill federal pledges of hardware like in the old days. Energia would supply the hardware, promised Semenov, but would work only on a commercial basis, like Boeing.

Semenov's speech on that fateful Saturday should have been a beautiful moment in Russian-American space relations. I knew what he was going to say and was, well, proud. Semenov, who spent his life wrapped in secret organizations and secret programs, was openly confessing to the delegation of Americans that his space industry was bankrupt. Not an easy admission. But at the same time, this Soviet patriot promised total cooperation if the United States would support him in this experiment called freemarket. With funding from the United States, he could build at competitive prices the critical components for NASA's space station.

But the blunt language and frank talk of money was exactly the message that Dailey and Goldin feared most. For them, Semenov was threatening blackmail, the anarchy of the Moscow streets being brought into the inner sanctum of an international cooperation now being discussed by presidents Bush and Yeltsin. What they chose to hear was something along the lines of "pay me or else." Dailey, who had worked in government his entire career, understood little of the internal difficulties facing NPO Energia and the Russian production lines. Dailey belittled Semenov as "desperately trying to negotiate directly with us to see if he could get money into his coffers."

The misunderstandings accelerated. Goldin and Semenov started arguing and soon it became a shouting match. Goldin yelled that he would never pay for hardware and Semenov better get on the reservation or be left behind. Semenov screamed back. The two proud and emotional men went at it. The lunch banquet was hardly touched, I understood. The American delegation stormed out and went directly to the small office of the Russian Space Agency. There, Yuri Koptev began his multi-year rapport with Dan Goldin, calmly assuring both Dailey and Goldin that yes indeed, Boris Yeltsin and the Russian government was prepared to cooperate with NASA. It was a self-described friendship that Dan Goldin never broke, while Koptev played the relationship with far greater sophistication.

Who was to blame?

Certainly Goldin for losing his temper in a meeting on foreign soil. And Semenov for not understanding how to speak with the American officials. From Semenov's perspective, the situation was straightforward. The Russian Federation had been in power for only six months. The Russian Space Agency was just four months old and still lacked the legal instruments for undertaking international business. Yuri Koptev lacked funding, nor did he enjoy the stature of Semenov.

Across the table Semenov saw two American officials that were neophytes in their own government. One had been working in the White House for four weeks. The other was an unknown figure in the American space community, just three months into his tenure as NASA administrator. Nor was this a new desire by Semenov. In 1991 Semenov was part of a Soviet delegation lead by Oleg Shishkin, a senior government official, that had met with Vice President Quayle and Dick Truly. The Soviet delegation had surprised the Americans by proposing a quasi commercial program in which the space shuttle would dock with the Soviet space station Mir. Quayle turned the suggestion over to the State Department where it died. But the proposal has all the fingerprints of Yuri Semenov. Pushing to integrate with America, pushing to tie the two programs together. Believing they needed one another to survive into better times, while wary of the poor track record of Russian politicians.

Dan Goldin would have been surprised to learn how Semenov fretted over the ramifications of Russia doing a deal with the United States and not being able to implement. Far from being the money grabber as depicted by Dailey, Semenov worried more about the price of failure. "It will be the end of cooperation for decades, maybe forever," he told me before the American delegation arrived in Moscow. "If we don't help each other today;

tomorrow there is no manned space program for either country."

This was the mindset of Semenov in receiving Brian Dailey and Dan Goldin. When Goldin responded to Semenov's blunt talk by screaming, it seemed to the General Director an act of unbelievable rudeness and stupidity. I saw Keller soon after the blow up. The man was literally still shaking. "I've never been so embarrassed in my life at what happened. Who the hell does Goldin think he is?" But it got worse. Next was the food fight over the Memorandum of Understanding.

The American delegation continued on its fact-finding trip, viewing a number of factories and organizations. Some of the facilities were viewed as possible help to NASA and others were of interest to the intelligence community, given their offensive weapon capabilities. The Russians were showing the Americans pretty much what was requested. It was yet another show of transparency. For veterans like Arnie Aldrich, the openness shown by Semenov was a shock. Aldrich recounted how during the time of Apollo-Soyuz they would meet their Soviet counterparts in rented facilities in the center of Moscow. Now, when he had arrived in March with Sam Keller, they were picked up in the Energia van and taken directly to the Energia facilities. "I've discovered that most of the guys we worked with during the Apollo-Soyuz program were really NPO Energia engineers." Aldrich explained that during Apollo-Soyuz the Russian engineers identified themselves only as being members of the "USSR Academy of Science." I'm not sure Dailey and Goldin understood that when Semenov opened his plant to the foreigners, it gave political cover for Energia's biggest subcontractor to do the same, and for their subcontractor and so on down the production line.

At the conclusion of the trip it was time to put into writing the conditions for moving forward. Dailey continued to misread the mindset of Semenov when he complained of "having serious problems with the Russians. They wanted to charge us a lot of money to hook up, and we didn't believe that since this was a government-to-government activity, that money should be involved, and it was the intention of the two presidents that something be put together that would be funded by their respective governments."

Noble. Traditional. But Semenov on the Russian side and Keller on the American side knew that given the empty Russian coffers it was a fairy tale.

The fight was over the wording of the agreement to be signed between Yuri Koptev and Dan Goldin. The incident began in the business center at the Radisson Hotel in Moscow, where the delegation was staying. Keller and his NASA team were writing their version. Gerald Musarra was in charge of what became known as the "White House" version. Two teams of a single American delegation producing two final terms and conditions for future cooperation. One involving a commercial foundation, with NASA paying for goods and services directly to the vendor, and the other relying on cooperation with the Kremlin. From everyone I heard there was a confrontation between the NASA team and the White House team with Musarra finally retreating to the government limousine to finish the document. As Dailey commented, "of course, the White House version won out."

The result was awful. The earlier Keller contract was to be altered so that it was not directly between NPO Energia and NASA, but rather between the Russian Space Agency and NASA. Goldin and Koptev signed this new agreement, leaving unchanged the funding and the contract requirements, but making Koptev's agency the responsible organization.

Goldin's Russian Policy

The change in policy was apparent immediately. In August a new NASA delegation was scheduled to visit Moscow. Sam Keller would not be going, or Arnie Aldrich. Instead, Dan Goldin chose a Marine, and former shuttle commander, named Brian O'Connor.

What a moment for O'Connor. Over the next year, with no experience working delicate international negotiations, O'Connor represented NASA in negotiating to implement the broad outlines of the space agreements that came down from the commission known as Gore-Chernomyrdin, for the vice-president and prime minister.

O'Connor was a solid guy who understood his limitations. He admitted that, "When I first got this assignment, I was a little intimidated by it, because I didn't have much experience with international partners or negotiating high-level agreements with other countries and that sort of thing. So I was all ears for people who had been through it before. There was a fellow named Sam Keller. I spent a lot of time with Sam Keller." O'Connor also spoke with Arnie Aldrich to better understand how to deal with the Russians.

The whole turn of events was perplexing and depressing to Semenov and Legostaev. For the Russians, continuity in negotiations is key, as it allows the two sides to develop trust and understanding. Not so with Goldin's NASA. Over the next several years the leading NASA negotiators were wheeled in and out. Each well-meaning negotiator arrived filled with misconceptions and frankly, just plain inexperience. Ryumin took the situation far worse than anyone else. As head of what was called "Mir-Shuttle" for the Russian side, he was forced to continually deal with new players that stuck blindly to a script being written by Dan Goldin and George Abbey. "Like negotiating with a computer, not a human," complained Legostaev. Ryumin was appalled at having to negotiate with Brian O'Connor, who, in his eyes, was both a novice negotiator and a dilettante astronaut, just slightly above a space tourist. The management of Energia respected, and preferred to negotiate with George Abbey, the head of Johnson Space Center. Abbey had his own power base in the Texas Congressional delegation, could influence the NASA administrator, cut corners when necessary and kept his word. At one heated meeting, lasting several hours, Abbey found a way to squeeze a few extra dollars out of another line item to keep Energia on track. This dark and brooding NASA official was sometimes hard to understand, but he shared with Semenov a dream that the damn space station was just an interim step to allow full cooperation on moving humans out into the solar system. However, Abbey was just one of many NASA interfaces for the Russians.

The procedure of having the negotiators rotate in and out may have assured complete loyalty to Dan Goldin, though even astronaut O'Connor was at times off-message. "Russians were going through quite a transition of their own," he later explained for the NASA Oral History Project. "They had a new outfit called NPO Energia, which was a semiprivate company which had spun out of their pure government operation, although it was the same people. They were also beginning the first days of their own Russian Space Agency. So they had sort of an equivalent to our Boeing, which was NPO Energia, and they had their equivalent to NASA, which was Russian Space Agency." The view of Energia as equal to an American contractor was a misunderstanding that caused so much trouble. From the American perspective, a space contractor saluted and obeyed the space agency. For Russians, it was the opposite.

On the Russian side the negotiators were unchanged and would remain so, despite the efforts of Goldin. Semenov refused to play second-fiddle to Koptev, fearing that the government official would continually bow to his own political pressures and promise far too much. Ryumin and Derechin and Legostaev would handle the top-end political and financial issues for the Russian Space Agency. Ryumin was the front line for Mir-Shuttle. For technical issues, it was usually an Energia engineer, with Victor Blagov of TsUP often involved on mission control issues. Koptev's Russian Space Agency simply lacked the expertise. They participated in the negotiations with NASA, but deferred to Energia on the operational decisions.

By the end of 1992 NASA announced that Sam Keller was retiring. He took with him NASA's understanding of how to work with the traditional giants of the Russian space industry. The agreements that followed with Russia in 1993 went far further, and faster, than I imagined possible. It ended up allowing NASA to finally realize the dream of a space station via cooperation with the Russians. Later, the cooperation saved that space station when the space shuttle Columbia's deadly re-entry over the skies of Texas grounded the space shuttle program for a second time, and Russian spacecraft kept the station operational.

Yet, more than once, while sitting in some hotel room down in Houston or at the Cape with Legostaev, we wondered what might have been if Sam Keller and his team had been allowed to remain with the program, and NASA had supported the economic reform efforts of NPO Energia.

Commercial relations continued at the insistence of Energia. Yuri Semenov and Bob Minor of Rockwell sealing the deal on the APAS docking system. Photo:Energia Ltd

Dan Goldin inspecting the docking hardware along with (left to right) Yuri Semenov, General Tom Stafford and Vladimir Syromiatnikov. Watching on the far right is Sam Keller. Photo: NASA

Chapter 10: NASA Buys Mir Services

Frank Morring is a solid reporter who for years covered the nuts and bolts of the industry through the dry pages of *Aerospace Daily*. It must have been the Christmas spirit, but at the end of 1993 he wrote an unusually colorful piece on our new office. "NPO Energia's new North American office," began the Focus feature, "reflects the Russian "scientific industrial corporation" in its shift from prestige pillar of the Soviet state to competitor for commercial space business wherever it can find it. Half furnished, with its files in moving boxes and no one to answer the phone because of a no-show temp, the three-floor facility on the outskirts of the tiny Old Town section of Alexandria, Va. is nevertheless a seat of high-stakes business negotiations.

"Perched on the stool left vacant by the missing receptionist, Alexander Derechin, NPO Energia's chief economist, goes over the fine points of negotiations for the $100 million-a-year NASA contract with the Russian Space Agency by telephone in Russian." The Focus articled captured the frantic pace that hit us, and how comfortable the Russians had become working out of the American office.

Much of 1993 was a dizzying time. I had worried when accepting Energia's funding that it could turn out there was little for us to do. And that cooperation would move forward in fits and starts. I needn't have worried on either front. There was a lot of work.

Faranetta worked with Energia on a very cool project known as Znamya, which was a reflective banner that was unfurled in space from the Mir, causing sunlight to beam down to the earth. Observers in northern parts of Europe and as far south as France saw a beam of light equal in intensity to several full moons. The idea of Znamya was to provide artificial sunlight to northern regions where inhabitants spent months in total darkness. It was a typical Energia project: innovative, using space to solve an environmental problem and short of cash. Faranetta tried to help, but in the end we received dozens of phone calls from people angry that Energia, they believed, was messing with nature. Never mind the complaints came from Americans comfortably living in our well-lit towns and cities. Corporate sponsorship became hard to find given the unexpected vocal opposition.

On the political front, newly elected Bill Clinton surprised everyone by retaining Dan Goldin as NASA Administrator. The new Administration also made the decision to disband the National Space Council, meaning that Dan Goldin was now the sole voice advising the executive branch on the civilian space program. Semenov took the survival of the NASA Administrator in stride. Semenov was a pragmatist who had spent his entire career dealing with powerful political enemies, whether in the Kremlin or from the other Russian organizations. One more powerful official at odds with his dream of commercially marketing the Mir space station and using the funds to move outwards to Mars was a bump in the road in return for the opportunity of breaking into the American market.

In March of 1993 Semenov and a core group of Energia engineers took a secret trip to Seattle, where Energia and Boeing officials mapped out a stunning program for a new station, one with contributions both from the Russians' planned Mir-2, and from NASA's Freedom. Boeing was now the single prime contractor for space station Freedom and was seeking some means to lower the cost of the program. At the same time, Boeing was eager to capture Clinton's support for the over-budgeted and unpopular space station. Working with the Russians on a new space station would accomplish both.

Boeing's Russian outreach was overseen by an official named Dick Grant. This always smiling, rarely discouraged executive understood Semenov as well as anyone in the aerospace business. One meeting stands out. In 1993 NASA still sought to justify the billions for space station as a means for achieving medical breakthroughs, even though Payload Systems had demonstrated less encouraging results. Boeing sought to fund a commercial pharmaceutical research project on the Mir to spur additional NASA support. Dick Grant and his team pushed through the negotiations for the crystal project in early 1993. The payload quickly flew and was expected to land in the autumn.

During a later corporate review between Energia and Boeing, the meeting dragged along on a half-dozen major problems. Finally, Dick asked if there were any other problems that needed to be discussed. Semenov hesitated and then looked straight at Grant. "I must tell you that with regret we cannot honor our promises regarding the crystal growth project." Another American executive might have blown a fuse. It was a pressure cooker time, as the presidents of Russia and America wanted to believe that the industries once geared to fight the Cold War could work efficiently together. Instead, Grant sincerely thanked Semenov for providing this information. "I know you are a man of your word, Yuri," Grant began. "So something must have changed for Energia to be in this situation."

Semenov shifted in his seat. He admitted that yes, something had changed. The unmanned spacecraft carrying the returned material was coming down later in the year than expected. Therefore it was no longer reasonable to assume the pharmaceutical crystals would be delivered immediately to Moscow, as promised. Again Dick thought it through. Everyone was silent, giving him time. "Yuri, I admit I'm confused. What's the difference in our contract terms because the capsule is returning later in the year?"

Semenov looked as if he was speaking from the dentist's chair. The proud General Director despised not keeping his word and here he had to bare his own internal problems to Boeing. "The capsule will come down in December. In December it is the darkest time of the year." Grant waited, still mystified, but now, like everyone else on the other side of the table, very intrigued. "In December daylight is fleeting. By tradition we have villagers who find our returned capsules. But in the dark of winter there is a greater chance of a local becoming lost or being attacked by wild animals. Therefore, the teams will look only in the limited hours of daylight. Instead of having the research package returned to Moscow in a day, as I promised, it could take two, maybe three days." Semenov stopped. Embarrassed.

Dick let out a quick laugh of relief that the issue was not political or technical. He again thanked Semenov for his honesty and added he expected nothing less. That their friendship was paramount to the business. Would a small donation of money to the village help motivate the local team? The Energia delegation breathed a sigh of relief. Semenov checked with Legostaev who whispered a few words of advice.

Within moments a modification was agreed. For Americans, the hesitancy of Energia was difficult to understand. It had to do with revealing to Boeing how Energia's internal business model had shifted from one of the traditional barter system to cold, hard cash. Cash was something Energia lacked, and very possibly the villagers were refusing to work until receiving more money than earlier negotiated. Inflation may have also been a problem. Basic commodities from Soyuz parts to bread and meat were shooting up in price within Russia, and for Energia to misjudge the cost of a service, including the finding of a capsule in the middle of the Russian winter, could be painful. I was present at a meeting where NASA officials laughed at the race by the Russian side to lock in a contract by the end of that calendar year. To the Americans, it was another sign of financial desperation by the Russians. Not understood was that after the Russian New Year holiday season, a key commodity used in the manufacture of the

Soyuz was expected to jump in price. This was something alien to NASA negotiators, and the Russians would rarely explain their own dilemma. Explanations of this sort were only possible if a personal relationship existed with the NASA negotiator, which was rare.

Grant could have responded angrily and accused Energia of not honoring its contract. In that case Semenov would have said nothing more. Instead, Grant handled it correctly, and Boeing reaped the rewards. In the second week of December the prime contractor was able to announce that a Boeing commercial crystal growth project had taken only eight months from the start of negotiations to being on the Mir to the return to earth. And, proudly stated a Boeing official, "we received the crystals in Moscow on the same day they landed in Siberia. We are very, very pleased."

That was Dick Grant and his Boeing team. Grant soon moved on and that was a shame.

In the second week of March, Semenov and the delegation returned from Seattle brimming with satisfaction. Semenov had found Boeing to be open to a radical restructuring, one which would eliminate Freedom in favor of a new station combining hardware from both nations. What was once considered quixotic was now within the grasp of realization.

Koptev was coming to Washington and the two men went to see Dan Goldin on March 16th. Goldin jumped at the suggestion of bringing together the planned space stations of both nations into one new station; so did the Clinton White House. In the months that followed a dramatic deal took shape. The first shoe to drop was in April, when NASA agreed to move the inclination of the space station to the Russian "highway " of 51.6 degrees above the equator, rather than the traditional American inclination of 28.8. This change would allow efficient use of the Russian launch vehicles.

Starting in July a plan was cobbled together calling for three phases lasting until 2001. "Phase One" was a series of shuttle flights to the Mir station, involving long duration stays by NASA astronauts on the Mir. These flights would take place from 1995-1997. Eventually seven NASA astronauts spent almost a total of 1000 days onboard the Mir space station, more by far than the total time in space of the space shuttle fleet.

NASA would pay roughly $400 million for the opportunity. There is an entire story involved in this dollar amount, one that underscores the miscommunication that plagued the discourse between the two sides. Briefly, very briefly, the Soviet organization Glavkosmos had signed a deal several years before to furnish both rocket engines and the underlying technology to the Indians. The contract was worth about $400 million. After great pressure exerted by the Clinton administration throughout 1993, including dangling in front of Boris Yeltsin the participation in NASA's new space station, a compromise was reached. Russia would send the engines, but not the technology involved. America would pay Russia the $400 million involved. It was a good deal for everyone except for Glavkosmos and the Indians. The General Director Alexander Dunayev fought a losing battle to stop the American-Russian deal.

I must have repeated the above brief explanation a dozen times in subsequent years to confused Congressmen, White House officials, and reporters. NASA negotiators spread the word that the Russians were "greedy," "hungry for money," and so on. From the perspective of space agency officials, NASA had arrived just in time to save the bankrupt Russian space program, and they should be damn grateful.

From the Russian perspective, they had been forced to end a commercial project with a long-time trading partner, in return for the same amount of funds from an America that was gaining billions of dollars worth of experience in exchange. So there was no reason to thank NASA. I showed whomever would look the dozens of published articles, mostly from European publications, that accurately depicted the history of the $400 million figure. That made the Russian's tough negotiating stance more understandable and more honorable, which I felt it was. But Energia's aggressive negotiating continued to cause its reputation with NASA to be dragged through the mud, hurting us with Congress and the financial communities. Many in the international media blamed Semenov, not Koptev, for the Russian problems implementing the new agreements. "Another sour spot is NASA's relationship with NPO Energia, the giant Russian company that dominates the country's manned space programme," was one typical lament, this from an Indian publication's December, 1993 year-end review. "(The Russian Space Agency's) weak position had slowed US-Russian planning for the scheduled docking of Mir in 1996, because Energia objected to it not being paid by the Russian agency." The reporter has Energia as the problem, instead of the inability of the Russian government to pay, or the refusal of NASA to recognize the problem and treat Energia as a commercial vendor.

After the initial Mir-Shuttle program, "Phase Two" would involve Russian contributions to a new space station that would be led by NASA. NASA suggested the new space station should be named Alpha, but Semenov hotly rejected the implications of this being the first, or alpha, station, saying it was an offense not only to the Russians, but also to the NASA Skylab. Even the White House weighed in. I was incredulous to learn that George Stephanopoulos, the media advisor to President Clinton, sent over a note to Goldin questioning why a change from the name Freedom was even necessary. Goldin, wisely, disregarded this bit of White House advice. "Why do you Americans so easily forget your own past?" Legostaev asked. As usual, I had no answer to this sort of question. The name chosen by bureaucratic default was the International Space Station.

After intense and difficult negotiations first in Crystal City, Virginia, just across from National Airport, and later in a horrible underground bunker at Johnson Space Center, it was agreed that the Russian Federation would supply the first module of the International Space Station, known as the Functional Cargo Block or FGB. Russia would also supply the second, core module for the space station, known as the Service Module. But the Service Module, which was based on the core section of the Mir station, would be under a commercial contract, and NASA would supply the funds. In addition, Koptev's Russian Space Agency promised to contribute the fuel and cargo supply missions via the Progress cargo ship, thus reducing the number of shuttle flights. NASA would supply the node attaching the FGB and the Service Module. This was all planned for 1997.

Koptev promised six to eight utilization and resupply flights per year. This allowed Goldin to announce a savings of at least one billion a year during the next four years of the Clinton administration. In August the name Freedom was officially dropped by Boeing. In October the deal was formally in place, ratified on the program level, then the agency level, and finally by Vice President Gore and Prime Minister Victor Chernomyrdin.

It was a huge victory for NASA and Dan Goldin, as the space agency was now viewed by the White House and the Kremlin as a key component of the warm relations between Russia and America. Congressional support would be more forthcoming, given the high-level international agreements. However, more than anyone else, the announcement was a startling vindication for Yuri Semenov, who since taking control of Energia in the late 1980's had pushed to integrate the space programs of the United States and Russia. What had seemed laughable in 1988 to just about everyone, was now a reality. Russia and America would build the new space station, with Energia as the prime contractor on the Russian side. And yes, as in the Sam Keller contract for the Soyuz lifeboat, NASA had agreed to pay for the use of the Mir, as would a commercial customer in any normal market.

But that ticking time bomb activated by Dan Goldin and Brian Dailey was still there. Firstly, the funds would flow to the small and weak Russian Space Agency rather than to the prime contractors. NASA had clung to the Cold War model of space, leaving the Russian government in the critical path, and expecting it to supply the first space station module, plus multiple annual flights of cargo and manned ships at no cost. That first station module would not be cheap, even using internal Russian accounting. Energia and Khrunichev probably required at least $60 million to build the module. The Russian Space Agency did not have that kind of budget. And with an annual inflation reaching 15%, every month of delay would considerably raise the cost.

Semenov immediately began warning of the inevitability of missing the future milestones. He became like an unwelcome relative. Same with me. I was sent to NASA four times over the following two years with the specific message that there were no funds to implement the Russian side of the agreement. My message was repeatedly dismissed by John Schumacher, the head of international relations for NASA. After one meeting Schumacher put his arm around my shoulder and explained that a message of this importance from an American was just not valid. After being told of Schumacher's brush off, Semenov sent Valery Ryumin to deliver the same warning. The candid message from the Russian head of the Mir-Shuttle program was also ignored by NASA. Dan Goldin and his team would then act surprised when each Russian International Space Station construction milestone was missed, yet the leaders of the nascent Russian industry could not have been clearer in their warnings.

Lobbying Congress

Energia Ltd. worked hard to explain our side of the story to Congress. Why it was good for America that the Russian organization Energia was taking the commercial path. How the Russian organization was restructuring itself to tap private sector funds to meet government obligations under the Gore-Chernomyrdin agreements. Throughout 1993 and 1994 we undertook a pretty aggressive campaign to tell Congress our admittedly complex story. Few could believe that the objective of Yuri Semenov for an American style market for space operations was receiving stiff resistance from NASA leadership, and that much of the funding for the Russian obligations under Phase Two was coming voluntarily from unrelated commercial projects of what was now called the Russian Space Company (RSC) Energia.

Sometimes the reaction was disbelief or even laughter. In 1993 Energia produced a television commercial on the Mir for a German firm-that, more than anything I could show in our presentations, demonstrated to the impatient politicians the commercial path being taken by Energia. So too the milk and soft drink commercials which were later filmed. Given that space station funding bills were passing in Congress by only a handful of votes, our outreach made a difference. I'd estimate we picked up between three and five votes from Congressmen supportive of Russian reforms.

None was bigger than the September 1993 switch of Senator Tom Harkin. As reported in an industry newsletter, "Senate rejects another attempt to kill the station. Noteworthy is that Sen. Tom Harkin, long-time station foe, has switched sides in the wake of the U.S.-Russia deal." From the floor of the Senate the Iowa Democrat stated his hope that the Russian involvement would reduce the chance of Russian space experts working for the likes of Saddam Hussein. The Senator then took on directly the arguments of Senator Bumpers, an opponent of Russian cooperation, who had earlier asked how NASA astronauts would feel about using Russian hardware.

"Well, as a matter of fact, Mr. President, they might feel pretty good because the joint agreement is with, of course NPO Energia of Kaliningrad. Under the regulations of the Russian Republic, NPO Energia of Kaliningrad has control of assets. Therefore, they must sign off on programs involving its' own assets.

"NPO Energia is the world's oldest and largest space organization. Like many Russian organizations it is being transformed into a commercial company."

This voting switch by a space station opponent was duly noted by a pleased Energia management. Legostaev called me into his office to compliment me on a job well done, as well as to better understand how these things took place in Washington. I explained that first I would brief the Congressional staffers using a two page fact sheet written with input from Alexander Derechin. Then, sometimes, I would have a moment with the Congressman or Senator.

Legostaev waited for more. I volunteered that Harkin was a long-time supporter of peacefully working with Russia. "That's nice," said the old Cold War warrior. I had no more to say. Then I realized what he expected to be discreetly told. "You understand, we would never pay a campaign contribution to any politician, Victor Pavlovich."

I'm not sure he believed me.

Some in Congress, whether new to space cooperation like Harkin, or veterans of NASA oversight, understood the situation far better than Dan Goldin. From the outset of the 1993 agreements, Congressman Jim Sensenbrenner, the Republican from Wisconsin, was vocal in his concerns regarding having the Russia government "in the critical path," as it was called. Joining him was Republican Dana Rohrabacher, the former surfer and speechwriter for Ronald Reagan, who also grasped quickly the flaws in the Goldin strategy. So too Democrat George Brown and Republican Bob Walker, both active in overseeing NASA. Their concerns grew louder as the Russian industry complained of insufficient governmental funding. Month after month, year after year, the FGB was late, which made all of the International Space Station late.

Each year from 1993 onwards Semenov wrote to the Russian Prime Minister politely informing him that there were no funds to meet the obligations to the Americans. Vice President Al Gore finally raised the issue in 1996 and was promised the funds would immediately be sent to Energia. They were not. I would say that it was in 1997 that a private sector solution was found. Russian banks advanced the funding for Russian space station obligations, with the loans backed by the central government. The FGB was launched in November of 1998. Not unusually late for a major new piece of space hardware, but late enough to have permanently tarred relations between a frustrated Congress and the Russian space industry, much to Semenov's frustration.

The iconic image of U.S.-Russian space cooperation which was taken over NASA objections. The picture was then released by NASA copyright free to the frustration of Energia. No one ever wondered who at NASA had taken this picture.

Photo:RSC Energia

Chapter 11: The Competition Virus

Part of my role within Energia was to explain in language Semenov could understand the psyche of the American political and space communities. What made NASA tick as an organization? What propelled Clinton to keep Goldin in his job despite his having served under George H.W. Bush? The questions, usually from Legostaev, could be specific or general. Legostaev startled me once by asking how Boeing could be the number one aviation company in the world but so disjointed when it came to the space program. He was right. The famed efficiencies of the manufacturer of the Boeing 777 were nowhere to be seen in the space program. "If you were a young American engineer," I finally offered, "and wished to advance within Boeing, and make good money, the aviation side of Boeing with its commercial dynamics and international marketplace, would be the right career choice over the government driven space side of the company". Legostaev thought this an illuminating answer and mused how wonderful it would be to work with the most competitive parts of the industrial landscape. Slowly, Energia's management had come to understand that our space program, with its dominance by the government agency NASA, was divorced from the famed efficiencies and vitality of America's markets and industries.

Sometimes Legostaev himself undertook some sleuthing to better understand America. One evening he was absent from the downtown Washington hotel where he and the delegation were staying. This was very, very unusual. During negotiations Legostaev was usually the den mother, his hotel room stocked with food and drink and always open to the others. But on this night no one knew where he had gone, so I waited in the lobby. About nine in the evening Legostaev came through the hotel entrance in his oldest and most faded blue jeans and leather jacket. He brightened when he saw me. "I have conducted an interesting experiment," he proudly disclosed. It turns out this respected space adviser had stood in front of the White House for several hours, asking those passing by for directions to the hotel. He wanted to see the reaction to his scruffy dress and "poor English" from average Americans. I told him he was lucky not to be arrested, but hoped he would share his findings.

"Yes, it is very interesting," he explained once we were in his hotel room and having some meat, cheese and a bit to drink. "You see, to my surprise, your black Americans were very nice to me. The nicest. They took the time to listen." He thought a little more. "Women were very patient. But the businessmen, they would not speak to me."

About six months later I found myself with a free evening in Moscow. Remembering Legostaev's experiment, I duplicated his behavior, only this time standing on the edge of Red Square.

The next day I managed to get onto his schedule. Legostaev greeted me unusually formally when I was ushered into his cavernous office by Svetlana, his secretary. Turned out I was interrupting a meeting for a new venture that Boeing was promoting. On the left wall was pasted at least ten yards of lined engineering paper. From left to right were hundreds of carefully drawn boxes along a multi-year axis. It was the milestone chart for the multibillion dollar Sea Launch project, which planned to launch commercial satellites from a modified Norwegian oil platform in the middle of the Pacific Ocean. The other space partners were the Yuzhnoy company, which was the Ukrainian manufacturer of the Zenit rocket, with Boeing providing the funding, limited hardware and the marketing.

Semenov had decided the planned project was so integral to the survival of Energia that with little warning he picked Legostaev as the first Energia project manager, so I had caught him on a bad day. Or maybe it was a bad year. He was reviewing the technical schedule with the key Russian and Ukrainian managers, and rather than using a computer he had handwritten with a pencil on blue lined paper the hundreds and hundreds of technical milestones that would need to be met before first launch. When I explained the reason for my visit he was incredulous and impatient for my results. "Your women were very nice," I quickly began. "Yes," he joked, "I know Russian women to be nice."

"But the businessmen were also kind, stopping to give me directions to the hotel, unlike the American businessmen." Then I explained that the Russian military men were rude, ignoring my question in my poor Russian or brushing me aside. We concluded that American businessmen and Russian military officials were equal in their perceived self-importance within their respective societies.

It was a game, but more than a game. Legostaev studied us-strove to understand our Congress, our space industry, our mood-while NASA officials from the top down were usually in the dark about the basics of the Russian mindset, their politics and the world swirling around their Russian colleagues.

The Green Badge Incident

In August of 1994 I was invited to attend a multi-national meeting at the Johnson Space Center. I forget now the reason for the meeting, but given the huge number of people involved, from NASA, the European Space Agency, U.S. aerospace contractors, Russian Space Agency and Energia, it could not have been anything too important. But the meeting unexpectedly provided an answer to a question that Victor and I had often discussed: what was it about Energia that Dan Goldin's NASA feared most? We wanted to grasp why NASA behaved so irrationally when it came to the commercial objectives of Energia.

The trouble began for me immediately. I showed up at the main security gate just off NASA Route 1 in Clear Lake. As usual, I gave the nice older Texas woman behind the laminated plastic counter my drivers' license as identification. She expertly rifled through the huge pile of waiting NASA guest badges until she found mine. It was a green badge.

I was shocked. In the years of visiting NASA centers all across the country I had always received a white visitor's badge. A green badge is well, for foreigners.

Foreigners who you don't trust, and have to be escorted because they are a perceived risk to national security. Those with green badges are sometimes escorted even into the bathrooms. The green badge reserved for me had, in bold black magic marker, a notice that I was restricted to two buildings, one called building 2, where the meeting was being held, and one called 4 South, where the Russian Space Agency had their own office.

I asked the woman to call Tom Cremins, who worked directly with George Abbey, the director of Johnson Space Center. It was Abbey who while at the National Space Council had agreed to go along with my request to have a letter on White House stationary wishing me well with my new job working with NPO Energia. Tom was coordinating this meeting and I figured he might straighten things out. But no surprise, Tom couldn't help.

The kindly NASA desk officer saw the look on my face. "Honey, it's just a green badge. Is that a problem?"

It's hard for me to say just why I took the issue so personally. I think it had to do with my constant frustration that NASA officials seemed hell bent against Energia. I took the badge from the now apprehensive woman and stepping out into the hot sun walked over to the meeting. Usually I had my way in these sorts of disputes. A few months before Alexci Krasnov of the Russian Space Agency had led a delegation to visit with John Schumacher in Washington. Alexei knew I had known John for years, and he surprisingly invited me to come along. When we got to NASA there was a problem: the meeting was restricted to government space officials, no private sector. Alexei began to apologize but in a loud voice I told him not to worry, I had my Russian Space Agency badge with me. Krasnov looked at me funny, as I handed to the NASA security guard a white plastic card. The card had my picture, a picture of the Milky Way Galaxy, my name in Russian, and in big Russian letters, it said "Casino Kosmos."

I explained to the NASA security guard that Kosmos is the Russian word for space (true) and that the first word was "Agency" (not true). A few weeks before I had tried to enter the recently opened casino in the lobby of the Hotel Cosmos. The huge security guard rebuffed me. I needed an identification card to enter, but in a manner of moments two pretty Russian women had taken my picture and a purring machine had spit out this laminated card with my image and that of the solar system whirling next to me. I was so impressed I put the casino card in my wallet. The NASA guard in Washington, DC was equally impressed and I was allowed into the meeting. John said nothing regarding my presence and Alexei couldn't stop laughing.

So as I headed into Building 2 I comforted myself by thinking at least I was one for two in disputes with the NASA bureaucracy.

Walking into the very large room, there was a round table that held maybe twenty or thirty people. I took my seat next to Energia's cosmonaut Sergei Krikalev. Krikalev is a muscularly compact cosmonaut, who was once a gymnast and looks it. He is the ultimate spaceman, comfortable within himself and able to float effortlessly in space like a dolphin through the water, while short-timers flounder helplessly in zero gravity. Krikalev was classified by NASA as a Russian Space Agency representative, but he was an Energia employee, who would be provided by Semenov to the Russian space agency for a fee. Everything Krikalev and the other cosmonauts did while on a mission was on a commercial basis. Spacewalks demanded a high fee, doing a research mission a far lower fee. And, on their return to Earth, as was traditional, the cosmonaut was to receive a brand new car. The concept of a completely commercial astronaut made NASA astronauts laugh. But why? These are professionals being asked to undertake services. Pay them based on the jobs being assigned.

An hour into the meeting I glanced over to my right and a badge pinned to Krikalev's jacket caught my eye. It was a white NASA badge. The sort reserved for Americans only. He was not subject to any restrictions. I stopped the meeting. I was told later the European Space Agency officials had never seen anything like my outburst. Given my badge, and where I was sitting, they had assumed I was Russian.

"How come Krikalev has a white badge and I have a goddamn foreign green badge," I sputtered. In the surprised silence it was Krikalev himself who answered that NASA has provided him with a white badge since he had flown on the space shuttle that past February. It was necessary, he said, that he comfortably be able to visit all facilities at Johnson.

Krikalev had been selected by Semenov and Koptev to be the first Russian to fly on the NASA space shuttle, as the kickoff to Phase One of the Shuttle-Mir agreement. At first, NASA insisted that he undergo the full multiyear astronaut training; this to a man who had already spent more than a year in space onboard the Mir. In fact, during his training the Johnson Space Center held a party celebrating one year of total shuttle time in orbit, meaning by adding together all of the 7 and 10 day shuttle missions the space shuttle fleet had now accumulated 365 days in space.

Krikalev confided to me in that deadpan style of his "I felt a little silly. I have more time in space than NASA space shuttle fleet." Krikalev also proved a smart businessman. On his return from the February space shuttle flight he went to Semenov and demanded an American car, not the traditional Russian car. The logic being he had flown on an American spacecraft. Maybe he deemed it payback for the miserable time he had spent on Mir in 1991 while Soviet politicians publicly debated selling the station to the highest bidder. Whatever the motivation, the reasoning was ingenuous, and Semenov caved in.

Krikalev's explanation only made me angrier. Looking over at the NASA official chairing the meeting, I tore into him. "You mean that NASA allows Sergei Krikalev, a Russian citizen, loyal to RSC Energia, loyal to the Russian Federation, trained in technology, to have full access to all NASA engineers and all the buildings while I, a loyal American, with no degree in technology, am restricted and must wear an international badge?" The Europeans and Russians thought it was pretty ironic.

The reason behind NASA's attitude towards Energia was now clear. It had to do with how uncomfortable the space agency was towards this new type of international cooperation based on market services and not international diplomacy. NASA feared the commercial threat of Semenov's vision of the space industry, with true competition for goods and services and foreign bidding on American space programs, more than technology transfer risks from training a Russian professional like Krikalev. The State Department worried about technology issues, but not NASA. Technology was the common language among the Russian and American engineers and between the cosmonauts and astronauts. But the new politics of commercialization was the virus that could bring down the whole NASA kingdom. As a representative of the greater perceived threat, NASA would give me a green badge and Krikalev, "only" a space engineer, a white badge. It would be exactly the opposite situation for visits to either the State Department or Commerce.

From then on when visiting the Johnson Space Center I received my green badge with pride.

The Russians continually surprised NASA and industry officials with both their openness and with their customs. At the Mir Symposium in 1993 the U.S. space industry came face to face with the once secret experts that had developed and operated the Mir space station. Photo:Energia Ltd

Chapter 12: With Semenov

The international cooperation soon extended out beyond NASA and the manned space program. In addition to the Sea Launch project, there was the formation of a company known as International Launch Services (ILS), which marketed the Russian Proton vehicle for communication satellites in a deal engineered by the Bush administration. Interestingly enough, ILS was led by Brian Dailey, who after leaving the National Space Council took charge of the Lockheed effort with Energia and Khrunichev. The cooperation was growing, but so too the cultural misunderstandings and slights, to us in Washington and to the Russians themselves.

An early meeting with Lockheed is worth recounting. We were meeting with the aerospace company in their Washington, DC offices in Crystal City. An agreement was reached to enter into an exploratory contract.

Energia directors in the traditional post-launch toast. When enjoyed at the Kennedy Space Center it was met by outrage from those believing it was a sign of the Russians inability to control their drinking. Photo: Jeffrey Manber

As the meeting wrapped up, a Lockheed official walked into the room and handed Semenov a check for $15,000. This was a down payment for the contract. We were stunned. I tried to imagine what would happen if Derechin handed the chairman of Lockheed a check during a Moscow meeting. But that's how the contractors saw the Russians-as a desperate community that only understood cash. In front of everyone, Semenov handed me the check, joking how my office had just earned a bonus. Nor was this an isolated incident. Semenov for months kept on his desk an envelope with $5,000 cash from a NASA contractor who was one of the "eyes and ears" for NASA in Moscow. To every visiting Russian delegation he would show the envelope. "See," he would say, "these Americans think they can buy us for the price of two good Italian suits."

Traveling with Semenov was an experience. He never relaxed, not for a moment. For a time, Energia also worked with the Loral Corporation using Russian rockets and satellites for the Globalstar constellation, a low earth orbit satellite constellation for phone and data communications. I accompanied him on several visits to Loral.

During one visit, Semenov met with Loral Chairman Bernie Schwartz in his New York office. The meeting was to begin promptly at the ungodly hour of 9:00 in the morning. Ungodly because there was little sleep the night before. There was first the obligatory meeting after we checked into the hotel, some eating and drinking in Legostaev's room, a little gossip, and no leaving until Semenov ended the night. As soon as we arrived in the hotel, Semenov would carefully dole out the room keys, deciding who had which room. There was a very strong pecking order and I was often at the bottom, or second to the bottom if Faranetta was around. It was not a bias against us for being Americans. It was simply that we were the rookies, the newest members of Semenov's inner circle. Sometimes I had to run out to get some food for the guys. If I was not in the room, Derechin would be sent. That's what rookies have to do.

Finally, at about 1:00 in the morning we were finished. I fell asleep as soon as my head hit the pillow, only to be awakened by the ringing phone. It was Semenov himself, wanting me to return. When he personally called, he would often try out his English. Slow and methodical.

"Do-You-Understand-Me?"

"Yes, Yuri Pavlovich."

"I-Am-Semenov."

"Yes, I know."

"Come".

I found Semenov and Legostaev still sitting around the coffee table. Semenov was puzzled about something that had happened the day before in Houston. We had finished a visit with George Abbey and had gone for lunch before heading to the airport. The delegation settled on Fuddruckers, the do-it-yourself hamburger joint. The experience was a hit. As is typical at Fuddruckers, first we were given a hamburger on a bun, and added the extras as we wished from the huge piles of lettuce, tomatoes, onions, melted cheese and pickles. Restaurants with salad bars or those with help-yourself portions did not yet exist in Moscow, that would come in a few years. So the concept of an open condiment section was a novel experience. Semenov wondered aloud about the cost of the meal. As I recall, it came to about $5 bucks a burger.

Now, at 2:00 in the morning in New York, the General Director was mulling over franchise economics 101. "How is it possible," asked Legostaev, "that the meat, the cheese, and the vegetables, could be only 180 rubles? Is this possible? We don't understand." Semenov grabbed Victor's arm. "The meat alone," he hastily explained, "in my kitchens, cost 250 rubles, plus the other ingredients. There is no way I could come close to that price for my workers. How can they sell all that food for so little money?

Pouring myself some Bloody Mary mix I led Semenov down the road of the cost efficiencies of outsourcing. I explained how Fuddruckers does nothing but buy the meat and condiments for hundreds of restaurants. Yet, Energia buys not only the materials for the Soyuz rockets, and the components for space station modules and communication satellites, but also runs the kitchens, assembles components for consumer products including artificial limbs, owns land for the dachas and so on. A small purchaser in many diverse markets. In America, I explained, there was a feeling that if you were not able to be among the best in a market, you let someone else do it. And, if not your core business, contract it out. "Some would argue," I told Semenov, "that Energia should let another company run the kitchen for your workers."

We worked through the issue for a little longer, discussing how European companies had the economic luxury to operate with one eye on social values, even if having to take a loss. Intrigued, Semenov asked for a report, including the annual amount of beef purchased by Fuddruckers, and their profit margins. He had come to understand how to use our economic transparency. Earlier in the year Lockheed had approached Energia to shoot an

IMAX film on the Mir space station. Ryumin went into the meeting assuming Energia would be paid for transporting the huge camera into space, as well as for assisting in the documentary. Lockheed instead insisted it should be a non-payment agreement. NASA Headquarters backed Lockheed, citing the public relations value. Ryumin was taken aback and stopped the discussions. Derechin urgently called to understand better the IMAX Company. "Is it like a Church, it is for the public good? Within a few weeks I presented a stack of documents on the publicly traded Canadian company. The documents detailed how much IMAX grossed per movie, their profit margins, how much their senior executives earned and so on. Ryumin used the supporting data in the next meeting, angrily questioning as to why IMAX should profit and not Energia. NASA philosophy won out and Energia transported the IMAX camera for next to nothing. The incident left a bitter taste in the mouths of Energia officials, that yet again because it was space, somehow money was taboo.

So Semenov knew that publicly traded companies like Fuddruckers had to reveal much of their costs, but fast food outlets were new to me. After some effort, I located a sympathetic stock analyst on Wall Street, and the report was dutifully faxed to Moscow. The impact of our conversation soon materialized. Some months later, back in Moscow, I was waiting for the usual Energia van to the factory. What arrived was a clean, white van, belonging to an outside transportation service. Semenov had begun shutting down departments no longer cost-efficient and handing them over to the new specialized firms sprouting up in Moscow. Many of the drivers were the same; the men had been hired by the company. The new vendor handled the insurance, which was a new concept in Russia, the vans, the repairs, the problems.

Semenov was ready to take on the Western practices that made sense to his company and his own sense of values. Not necessarily American, but Western. The General Director could be sentimental and traditional, but had no fear of change. He didn't stop with the vans. Semenov took a hard look at many of the traditional services, and demanded each be justified in light of the changing marketplace. If possible, shut them down and outsource the service to a specialized company.

Growing the Relationship

Given the late discussions in New York, the nine o'clock appointment with Loral arrived far too early. Semenov was always prompt, so the breakfast of leftover tomatoes, cold cuts, cucumbers and cheese began at 7:30 in Legostaev's room. Waiting for us when we reached the midtown office was the chairman and two of his senior advisors. Schwartz was about to begin the meeting when he heard me whispering to one of the advisors. His eagle-like eyes bore into me. "Who are you?" Demanded the Loral chairman. Just moments before Schwartz had pressed a button which slowly lowered a transparent mesh of metal fabric over the conference room windows, preventing any electronic eavesdropping. So his sudden question possessed an unexpected power, as if I was a caught-out interloper. "Are you the translator?"

I froze. No one had ever asked who I was. It was a really, really good question. Legostaev answered. "This man is Manber. He is one of us" It was a statement of fact expressed in a monotone. A little more was then provided. "He is our bridge." Schwartz was very confused. Legostaev repeated it with no explanation. "Manber is our bridge."

That was that. I had found myself. I was their bridge.

It seemed as good a job description as any. Job titles were not that critical, especially in the beginning, when we were working part-time and on a three-month funding leash, unsure whether the Sam Keller cooperation would take long-term roots, or, more importantly, whether we could produce results for Energia. Our relationship to Energia took a major leap in the middle of 1993. Probably much progress had already been made, because we started, from the Russian perspective, from just about zero trust. Soon after the first funds arrived, Artemov called the newly opened office. Faranetta answered the phone. A few days later, he called again. This time I answered. Boris expressed surprise. "If we sent funds to Russians working in another country, we would never hear from them again."

The first strengthening of our relationship came principally from our hosting a symposium in July of 1993 in Washington, DC devoted to the Mir space station. The timing was ideal, just as the political situation moved towards greater cooperation and NASA and Congress wanted increased transparency. *Mir Space Station, a Technical Overview,* was a huge event, taking up months of preparation and required all of our funds. Never before had the working engineers of Energia come en masse to the States, or been so readily available for questions.

We were on edge whether they would even arrive on time. The office couldn't afford to keep that large a delegation in hotels for too long, so we cut it close. It required a tight connection in London on Aeroflot and I just prayed. Somehow the flight was on time and a dozen senior Energia officials arrived at Dulles. Also arriving into Dulles airport was Sergei Krikalev, though he was then still in training for his February Shuttle flight. NASA had refused to release Krikalev when we asked for his presence as a speaker. Semenov went ballistic. The resulting compromise led to another first in American-Russian relations. As part of his training, Krikalev flew a NASA T-38 training jet from Ellington Air Force Base outside of Houston to Dulles, becoming the first Russian to fly an unescorted T-38 flight across the United States.

The day the men arrived we had another one of those moments that defined the growing relationship. We planned a picnic for the delegation, and had carefully laid out on the tables the meat for Shashlyk, the Georgian shish kabob, along with the tomatoes and cucumbers and sauces and drinks. When Semenov came upon the picnic area he stopped cold. "Where's the food?

In the shopping bags on the picnic tables I explained.

It was a major faux pas. The guest always arrives to a table filled with food. I knew that. It was true in every traditional restaurant and every home I had been in. Time to think fast. Very fast. "You are not a guest here, Yuri Pavlovich. Nor your men. This is your picnic, we are your colleagues. Roll up your sleeves. Have a beer and let's get to work." He let me get away with that, which was noted by the entire delegation. Soon the men were singing the old songs often sung by a delegation during a time of enjoyment or before going into negotiations. It must have seemed very strange to Virginia picnickers walking past.

The symposium was a success on all levels. NASA Headquarters had spitefully sent out a note prohibiting contractors from using NASA funds to pay for attending, which would have killed us. About a week before we had maybe fifty attendees. Good, but not good enough. Then someone at NASA relented, and hundreds showed up. For the average American contractor or NASA official, the event was an eye opener. There was no counterpart in the NASA space program to Leonid Gorshkov, the director of development of the Mir, nor Oleg Mitichkin, the expert on growth of crystals in microgravity, or Edward Grigorov the director of life support, or Vladimir Branets, the director of control and navigation, or Vladimir Syromiatnikov, the head of large deployable structures or Oleg Lebedev, the head of remote sensing. Energia's experts were the space programs equal to the Murderers Row of the 1927 Yankees, these men were the space program equivalent to the likes of Earle Combs, Mark Koenig, Babe Ruth and

Lou Gehrig. There simply was no equal in the NASA structure. In fact, it was not even encouraged. Semenov and his advisers were shocked to learn there was no "father" of space station Freedom, or "head" of the space shuttle. Engineers like Arnie Aldrich were rotated in and out. Semenov and Legostaev believed this to be the single greatest flaw in the structure of NASA.

It was not just Russians speaking about their Mir experiences. We had the French cosmonaut Jean-Loup Chrétien in his bright red sports jacket speaking about cosmonaut training. Sam Keller, having settled with the East-West Institute of Maryland, gave the introductory talk. Anthony Arrott, now consulting with Arthur D. Little spoke, as did Bob Reinshaw, still with Payload Systems. Brian Dailey was representing Lockheed, and two Europeans with Mir experience, Wolfgang Nellessen and Udo Pollvogt, also participated.

Two days before the symposium began I got a phone call from Tom Cremins. George Abbey wanted all the men to begin negotiations in Crystal City immediately. "Forget the symposium," laughed Cremins. I wasn't laughing. I called everyone I knew at NASA, up to and including Abbey. Given we were discussing a wait of just a few days, he relented. As soon as the event ended, a NASA bus transferred everyone for the start of the negotiations that led to the Shuttle-Mir agreement.

The night the delegation moved to Crystal City, right next to the Pentagon, Legostaev and Semenov came out to our office in Dulles. Artemov had suggested we prepare a new budget, in the assumption that the symposium would be a success. The Russians may not have understood all that we did, but they could see the symposium results. We had progressed far from the blue marketing book of two years before.

We prepared three budgets, starting with continuation of the current arrangement to full-up American representation. To my surprise, Semenov selected the largest. The agreement was signed that very evening. From this time onwards, Legostaev was not as involved, nor Artemov in our operations. Energia Ltd. became the responsibility of Alexander Derechin, who was now the head of the Economics Department. Derechin and I reviewed the new budget line by line, and saved negotiating my salary for the end. We reached several impasses in the discussions before finally reaching an agreement. "Whatever you do," warned Derechin, "don't ever tell Semenov your salary."

This seemed a little strange. Derechin explained that Semenov came from a world where the company supplies a car, a country home, medical insurance and the like. He wouldn't understand my salary in that sense. I took his advice for the next year, but then it happened. In Washington for NASA discussions, Semenov, unprepared for a cold snap in the weather, wanted to buy a hat. We went off, alone, while Derechin and Legostaev stayed at NASA.

After finding a suitable cap we stopped for some coffee. My Russian language skills were poor and Semenov's English worse. But Semenov interrupted the silence by asking how much I earned from Energia. At first I pretended not to understand, but still he insisted. Thinking through the options, there seemed little wiggle room with such a direct question. I told him that his advisers had told me never, never to reveal the number to anyone. It was a great secret. Semenov nodded sternly.

Fearing an impending execution, I wrote the number down on the napkin. Then I wrote out the number underneath in block letters. I handed him the napkin, and with a quiver in my voice, reminded him it was a secret. The General Director looked at the napkin. Then his huge hands folded the paper napkin once, then twice, then once again, before he popped the paper ball into his mouth, chewing until swallowing away my salary level. This was a man who knew how to keep a secret.

Salaries seemed a very sensitive discussion. One of our first visitors after we moved into the new Old Town office was Yuri Koptev. He sat in our conference room and reeled off a dozen questions. The cost of electricity. The cost for insurance. The amount budgeted for the furniture. The amount paid for the office and the details of the mortgage from the bank. Finally he stopped.

"Aren't you going to ask about our salaries?" I invited.

"Salaries are sacred. That I don't ask."

The Disruptor at Cape Canaveral

With the onset of Shuttle-Mir cooperation, Yuri Semenov made the journey to Florida to witness the first of several shuttle launches. He would arrive at Cape Canaveral with a delegation of senior advisors, including Energia board members and top program managers. The make-up of the group reflected the wonder that an improbable dream was being realized. Their space station, having narrowly escaped the budgetary death of the

Buran space shuttle, was now the pivotal center for the American space program. Semenov wanted his closest supporters to experience the moment, and, at the same time, needed the key engineers close by in the event of an unforeseen crisis.

There was no more jarring component for the well-oiled NASA pre-launch public relations machine than the Semenov delegation dropping into Cape Canaveral. The Russian's deeply held notions of ownership, pride and tradition went against the collective governmental experiences of those planning the shuttle launches. Every NASA representative, from the driver of our bus to the Mission Control launch director was forced to confront a situation never even contemplated; that the U.S. space shuttle sitting out there on the launch pad was programmatically and politically tied to a space station that legally or by fiat, belonged to the Russian company Energia.

The General Director didn't help himself in the eyes of the Americans. He behaved at the Cape much like Peter the Great inspecting the troops of a new political ally. NASA at first treated the Energia delegation no different than other visiting foreign contractors. A delegation to be politely hosted, but one certainly not to interfere with the NASA "show" in any manner. That was a mistake brushed aside pretty quickly. Semenov's delegation would travel in their own van, not sharing a bus ride with other visitors. That decision alone was a huge, huge bureaucratic struggle. The General Director would have access to "his" cosmonauts, in a manner consistent with Russian tradition, not NASA policy. Nor would the space shuttle be launched without the final green light from Valery Ryumin, after consulting with TsUP in Kaliningrad and finally his boss Yuri Semenov. Incredulously, then begrudgingly, and occasionally with appropriate grace, NASA relented. But more than once a NASA shuttle team member or public affairs officer professed feeling soiled by having to work on a program so identified with this "space peddler," as one suntanned public affairs officer called Semenov. There were bruised feelings percolating throughout NASA that somehow Semenov was responsible for the $400 million to be paid for using Mir. Rumors flew that the money was ending up with the mafia, or lavish homes for key officials. Articles were written about the supposed corruption of the Russian side with an unexpected vengeance, as if the injection of money into NASA's international programs poised a threat in and of itself.

The post-launch Shuttle celebration was one of the traditions that Dan Goldin accepted better than the NASA troops. STS-71 carrying Sergey

Krikalev was the first shuttle to usher in the new era, referred to by NASA as Phase One. Moments after Atlantis safely blasted into orbit, the Russians took out the carefully stored vodka, cups and juice from our van and toasted, with great relief, the successful launch. Energia worried about shuttle reliability, so the post-launch toast was more emotional than usual. What an uproar that swept through the NASA space community. "The drunkards in broad daylight swilling down their vodka!" The drinking right at the viewing stand solidified many prejudices regarding the Russians and their rumored drinking habits. Somehow, someone may have gotten to Goldin and explained. I know I later spoke about the tradition to John Schumacher.

For a later flight, I think it was STS-72, which carried Vladimir Syromiatnikov's docking unit to the Mir, Goldin shocked the industry by joining Semenov and Koptev in the celebratory post-launch toast. The sight of Goldin and Semenov standing together in the bright Florida sun, toasting with vodka-filled plastic cups while NASA aides nervously clucked about the possible public relations repercussions, was probably the finest moment between the two men. For Semenov, the space business was about working every day on the edge of what humans can achieve. There will be failure and when the space gods have been kind, you celebrate.

The image from this era of cooperation is without doubt the stunning photograph of the Mir and the Atlantis space shuttle docked together in the void of space; with over two hundreds of tons of space hardware delicately mated to one another. Despite the beauty, the photograph is yet another reminder of NASA's inability to understand Energia's reliance on the private marketplace.

The idea of taking the picture was the brainchild of an Energia engineer, and it quickly worked its way up the chain of command to Semenov, who approved immediately. It was the sort of big gesture that appealed to the General Director. NASA was aghast. Energia made plans to go ahead anyway, leaving NASA to plead not to have another manned capsule flying near Atlantis.

On July 4th, 1995, Cosmonauts Anatoliy Solovyev and Nikolai Budarin strapped into a waiting Soyuz and pulled away from the Mir. With a few short bursts of the thrusters, a safe distance was reached, from which the cosmonauts captured the majesty of the historic docking of former enemies embarking on a new era in space exploration. *The New York Times*' Bill Broad described the Soyuz as "Moving with the delicacy of a dancer, the relatively tiny Soyuz pulled back about 200 feet to photograph for engineers and posterity the giant East-West structure against the backdrop of outer space."

It was a source of great pride to everyone at Energia. What followed next was vintage NASA. The space agency released the picture, which was provided by Energia, for worldwide distribution. Overnight the photograph became copyright-free. As I write this, I've stopped to look up whether the image exists on the open source Wikipedia site. It does. And the text states, regarding the photograph, "This file is in the public domain because it was created by NASA." I wonder who the wonderful Wikipedia contributors think took the picture? It was so frustrating. My complaints to NASA fell on deaf ears; I suppose they thought me a "space peddler" as well. Semenov believed that photograph should have been protected as an Energia release. I completely agreed. We tried to monetize the Energia photographs of that period, but the attitude in the West was that because it was a space image, it was copyright free.

Some time afterwards, I gave an interview on the New York radio station WBAI, which is a progressive station known for its left of center views. When asked by the reporter how it was an American businessman could work for the Russian space industry, I replied that in all seriousness, and sadness, that if you want to work with a socialistic space program, go work for NASA. If you want to work for the capitalists in space, you must go work with the Russians.

Chapter 13: The 50th Anniversary Bash

Nothing says more about the status of Energia within the Russian space landscape than the birthday party thrown to honor itself. It was a theatrical space-themed bash mimicking one of those celebratory scenes from the *Star Wars* blockbusters, where hundreds of victorious space explorers return from far-flung planets and solar systems for a victory celebration. It was August of 1996 and the company was now known as RSC Energia, which since 1994 had formally transformed into a private shareholding company.

Energia was now owned partly by the government and a majority by shareholders. The transformation was a messy affair, in part because of the opaque regulations in Russia, and in part because of the immaturity of the Russian securities market. Each senior manager received 100 shares of the only one million shares released. In Washington, there was scornful laugh-

Yuri Semenov, Christopher Faranetta, Nikolai Zelenschikov, the author and Victor Legostaev during the huge two day 50th anniversary celebration for Energia. It was during this event that Semenov revealed the first ever public organization chart for Energia. Included were the author and the American office. For some within the Russian space program both the chart and the inclusion of a foreigner was a step too far. NASA never understood the pressures that Semenov felt from within his organization not to cooperate with the Americans and not to reform along Western economic principals. Photo: RSC Energia

ter at the thought that we had been given stock in Energia. The stock would soon rise, however, from $5 a share to more than $400 a share before collapsing back down in the 1998 Russian ruble collapse. What was silly was that at the highest price, the valuation of Energia was at the ridiculous level of only several hundred million dollars. Never was there a more urgent need for sophisticated input from the West to assist Semenov on restructuring. He went halfway and froze. Equity was a potent tool he didn't fully understand.

It was a good time for the "collective," as Semenov and other older officials still referred to the organization. The aerospace analyst Wolfgang Demisch had recently declared RSC Energia the strongest space company in the world. The problems with the Mir were in the future, and only the bankrupt Russian government and NASA's insistence to rely on the Kremlin for Russian hardware, marred the revenue potential.

Certainly Energia enjoyed the richest history of any space organization or corporation, and fifty years of survival was a great reason to celebrate. The commemoration opened with an all day concert in downtown Moscow, which lasted far into the night. In the first row were the senior Energia officials and distinguished guests. Sitting right in the middle of the first row was 87-old Boris Chertok, who worked with Sergei Korolev in the 1940's when the captured Germans were brought back to the Soviet Union. Chertok still worked three days a week and had changed little, except for the sneakers he often wore for comfort. Chris and I were in the second row, along with the personal guests of Semenov.

The concert opened with traditional Russian folk entertainment, followed by the surprise appearance of Yosif Kobzon, popularly known as the Russian Frank Sinatra. The older women in the audience swooned as his baritone voice delivered the popular ballads of his career. Semenov was on the stage the whole evening, doing a pretty good imitation of an Ed Sullivan host in one of those old variety television shows. He introduced the groups, bantered with the stars and at one point acknowledged our presence to the 3,000 attendees. The men in the front row were made of some pretty strong stuff. No one moved after one hour, or two hours or three hours. The music then turned loud and raucous, supplied by the hottest new rock groups in Russia, to the delight of the younger workers high in the balcony. The old men didn't budge. Not after four or five hours. I was fine. Sitting next to me was a well known and very pretty Russian actress, who was subjected to constant close-ups from a roving television camera. Finally, mercifully, around midnight, the concert came to an end.

The next day began with a group photo of all the Energia officers, past and present, before a huge Energia rocket. Then we moved into one of the great halls on the Energia property. Before us science-fiction became reality as hundreds of contractors, politicians, government officials and workers associated with the Russian space program swept into the huge hall to voice their support for Energia. Chris and I were included in the parade; we arrived with a letter from the Mayor of Alexandria, Virginia, home to our office, congratulating Energia on its 50th anniversary. Once the letter was presented to Semenov, Legostaev allowed us to stand with him.

It was a show never to be forgotten. Delegation after delegation would present themselves to Semenov, and then many also paid their respects directly to Legostaev. One group of representatives included men with long beards and the knitted hats often worn by Muslims. Each of the men stepped up and after kissing Legostaev would proclaim something like "we, from the collective in the southern Urals, who have supported you since the time of Korolev, pledge today our continued obedience." Legostaev would give a quick acknowledgment and the men retreated, to be replaced by yet another delegation from some corner of the former Soviet Union. One group of three men with simple and old styled dark business suits chose to ignore the recent independence of their country. "We, from the Ukraine, pledge our everlasting devotion to the collective, and will continue to supply the valves for the Soyuz engine."

Suddenly Boris Artemov gripped my arm. "You must not see this or tell anyone." Coming into the main hall were several large men who were clearly bodyguards. Then in marched Oleg Baklanov, chief of the Military-Industrial Commission. Baklanov was one of the organizers of the failed August 1991 coup. There was a murmur from the audience as he paid his respects to Semenov. Soon after I had met with Semenov in Montreal I first heard the rumors that Semenov had supplied the plane that flew the coup plotters to Gorbachev's summer retreat, where they urged the Soviet leader to resign. Whether he did or didn't I don't know, but some of the older men of Energia spoke publicly that Semenov had, indeed, supported the failed coup.

After the greetings from Baklanov, dozens and dozens of cosmonauts swept into the hall, to the deep-throated cheering of the large assembly. I had never seen so many space travelers in one place. Smartly dressed, many flashing medals of heroism, the men were of all ages, reflecting those who had taken the first steps into space to those who were privileged to have lived aboard the Salyut series of space stations and later the Mir. A few rookies represented the unknown future. Too few.

The rows of cosmonauts stopped before Semenov, shouting out their devotion. Cries of approval echoed back from the hundreds in the hall and for a brief moment I saw what I had only before felt. Far more than possible in America, I was seeing a community of space explorers, not a program. These workers, whether they were the regional leaders of subcontracting organizations or the major builders of the rockets and space station modules, or the spacemen, all gave their devotion to Energia. Not to the organization now known legally as RSC Energia or earlier as NPO Energia or even as Department 88. Not to Semenov the man, but to something far deeper. I felt privileged not only to bear witness to the event but to do so while there still existed an unbroken line from the flight of Vostok 1 and Yuri Gagarin to the new era of Sergey Krikalev and the NASA cooperation.

After the welcoming ceremony, we all moved to lunch. Chris and I were at a table for the invited foreigners. There were several Japanese customers, one official from a European space program and a NASA representative. During the lunch a small woman, looking in size and spirit much like Barbara Mikulski, began a fiery oration, holding the room spellbound. She was, I knew, a powerful Duma politician. The NASA official became furious. He knew little about the space program, as his background was intelligence. We were enjoying an informative conversation on the history of Energia, and its dispute with the Russian Space Agency, until the woman began her speech. "This woman is a communist," he shouted over the din. "Is Energia a communist organization?" I explained that the Duma was dominated by the Communists, and Energia was a centrist organization, one that had recently privatized. He wouldn't hear any of this. "This is a communist supported organization," he sputtered with indignation.

After lunch Legostaev approached me with the surprising news that Semenov would see me. He was greeting senior industry officials and politicians, so the invitation was unexpected. After a long wait in the crowded outside hall Legostaev ushered me into the conference room and back into Semenov's private office. Semenov greeted me warmly. I had last been alone with him, in his private office, just two months before. We had an embarrassing problem that Semenov had wanted to personally explain. Our payment for that quarter had failed to arrive. It was the first time Energia was this late with sending the office funds. In Washington, our reliance on Energia for funding was the butt of jokes; I was asked constantly from NASA, from contractors and Congressional staffers whether we were paid in vodka, or Russian rubles, whether we were paid at all, or whether we were paid with "suitcases filled with cash," as one senior NASA official warned an Administrator-attended meeting. The reality was that the home office was no better or worse than any home office.

Semenov laid bare for me the reality of Russian politics, circa early 1996. Boris Yeltsin was in a tough reelection fight against the Communists, and all of Russian industry was being asked to pony up to pay for the election campaign. Semenov rubbed his face with worry. "I have given a huge amount of money to support this president, money we planned for our operations. All of us in the industry were asked, and we are doing so." Semenov explained that all departments would receive their payments, but just late. "What happens if the Communists win?" I decided to ask. Semenov stared deeply into me. "The Communists are gone from this country forever."

On this day marking the 50th anniversary of Energia, the mood was far brighter. Semenov had a gift for me. He reached into a box under his desk and handed me a massive coffee-table book. It was more than six hundred pages devoted to the history of Energia. We had heard about this book for much of the past year. Semenov had personally poured over the text, deciding what to include from the secret archives and what to emphasize. It was a treasure trove of material. Long classified documents on secret Soviet programs. Never seen documents from Stalin, Khrushchev and Brezhnev. It was a concrete display of transparency by the old guard. Even the inside covers of the book held meaning. Each displayed an aerial view of the Energia territory. Overhead photographs of the manufacturing facilities were still a novel sight.

Semenov autographed the book and handed it over. But there was something else. Legostaev was grinning. "The General Director has given you a bigger gift. You are in our book." The two men then together rifled through the massive volume, looking for a particular page. The page being sought was an organization chart of the company. Yuri Semenov had finally done it. Five years after our meeting in Montreal, three years after allowing his experts to speak at the American symposium, two years after privatizing, the first publicly available organization chart of this once fiercely secret organization had been published. Semenov took his big thumb and pounded down on one of the boxes. Now I understood, my name was included. How strange to see a black and white acknowledgment of the strange fact that I was working for the Russian space effort. The chart showed me as heading the representational office in America, reporting to Alexander Derechin, who in turn reported to Victor Legostaev, who reported to Semenov. I accepted on behalf of myself and Chris Faranetta.

It was an unexpected gesture, but not one that all approved. Occasionally I ate in Semenov's private kitchen, along with the so-called retired advisors who had once actively run the Collective. Sometime later one was very angry at my inclusion in the book, and told me so. "This is reform far too far, to have a foreigner in our organization," said the former Energia manager, now in his 70's. Nonetheless, he didn't seem to mind eating soup with me.

That night, back in the hotel, I discovered one more surprise contained in the organization chart. The chart showed six columns of operational departments reporting to Yuri Semenov. Also reporting to Semenov were three advisory councils, including that of the general constructors. There were two boxes above Semenov. The first was the board of directors. But one more box was supreme. This meant that the chart indicated that all of the operational programs, the board of directors and Semenov reported to an even higher authority. That box represented the shareholders of RSC Energia.

I thought of the delegations from all across the former Soviet Union who had earlier that day shown their allegiance to Energia. These organizations were looking for Semenov to lead them out of the economic chaos back into stability. He took his leadership role very seriously and carefully chose a path of market reforms, not just for the present day, but also as a model for Russia moving forward. Far more of the Russian aerospace community would have followed, if NASA had not chosen to strengthen the Kremlin at the expense of these emerging commercial organizations.

Chapter 14: Fear Beneath the Petrified Salami

It is late in the evening and I am staring into the refrigerator looking for anything edible. It seems upon careful scrutiny that the small refrigerator in my Washington office holds on the shelves its own personal history of international space cooperation. There is a half-empty bottle of Bloody Mary mix-a favorite of Semenov-but only when he was in America. There are one or two cut up and very dried lemons, good for straight shots of vodka for those not having the taste for a Bloody Mary. There was a brown bottle of some mysterious drink from Siberia that no one has had the courage to open or the desire to throw out. We will get around to that someday. On the top shelf there is some hard salami; it was hard when fresh and now petrified. A gift from a visiting delegation of Russian government offi-

Despite the display of flags, NASA was considered by other space agencies as ill-prepared for handling delicate international diplomacy. Here NASA astronaut Bill Shepherd is surrounded by his Russian crewmates (left) Sergei Krikalev and Yuri Gidzenko. Shepherd's designation as the first crew commander for the International-al Space Station offended many in Russia and showed NASA's tin ear for international relations. Photo: NASA

cials, since Russians never drink without also eating. And lying on its side is an unopened bottle of vodka called "the pride of Russia." The label had a color picture of the Buran, the Russian space shuttle.

Oh, and on the bottom shelf of the refrigerator is a small box containing more than $100,000 worth of sophisticated electronic components ready for inclusion in a manned spacecraft or satellite or space station module. The story of the components is a perfect example of what has gone wrong with international space cooperation between NASA and the Russians.

Operationally, NASA never really came to grips with being in bed with the Russians on Mir. Seven NASA astronauts lived on the space station with results ranging from the routine to the catastrophic. There was the July, 1997 accident with an incoming Progress cargo ship that ripped apart the Spektr science module. This was the module where the NASA astronauts lived and performed their research. Not since Apollo 13 had a NASA astronaut confronted such a dangerous situation. The crew of Michael Foale, Mir commander Vasily Tsibliyev, and flight engineer Alexander Lazutkin endured a life-threatening crisis as the men fought to stabilize the station. The accident threatened to also rip apart the cooperation between the two countries. Dan Goldin faced enormous pressure from Congress and many of the astronauts to pull out of the crippled space station. I had never seen Legostaev so angry and worried. Angry that the Americans might be fair-weather partners. All of Russia felt the same way. "Would you turn back half way to Mars if we lost a ship," was a common lament.

Legostaev called regularly to find out what Dan Goldin was thinking, and also burned the lines with George Abbey and NASA's Shuttle-Mir director, Frank Culbertson. After appointing a commission to look into whether the space station was safe, Dan Goldin made a surprising decision. He elected to stay the course and keep to the planned rotation, allowing David Wolf to become the sixth, and second to last, NASA astronaut to live on Mir. Had something gone wrong during the four month stay of Wolf, or the subsequent almost five months for Andrew Thomas, it would have been a national scandal and the blame squarely in the lap of Goldin. "A strong man. Strange, but strong," was Legostaev's relieved re-evaluation of the NASA Administrator.

The space station soon descended into a punching bag for late night comedians. Dave Letterman weighed in with the "top ten complaints by NASA astronauts on the Russian Space Station", including:

- Its powered by a donkey on a treadmill;

- There ain't nothing messier than zero-gravity borscht;

- Ever since accident, they can't shut off the left turn signal;

And the number one complaint by astronauts living aboard the Russian Space Station:

- The damn thing smells like cabbage.

The scariest moment outside of the public eye came during the 1997 mission of Jerry Linenger. The Russian psychologists concluded their routine pre-mission evaluation of Linenger with the shocking conclusion that he should not undertake a long-duration mission. The medical conclusion was accepted by the Russian Space Agency and Energia. Derechin and Ryumin fought hard with NASA Phase One officials. Derechin swore to me that there was no politics, it was a straight decision based on decades of medical experience. NASA took umbrage and ignored the warnings, with the result being Linenger spent the most miserable and difficult four months of his life. At times he stopped communicating with the crew and ignored his team on the ground. In hindsight the Russian screening procedure was correct. "For a two week mission," explained the director of the Institute of Biomedical Problems, "just about any normal person can adapt. Longer than a few weeks it is a serious issue of internal training. Linenger lacked the right skills."

Even in the best of times the cooperation was strained. Few astronauts sought to adapt to the Russian system. NASA personnel kept to themselves in their own Texas-style suburban townhouses at Star City, ate their own foods and complained about just about everything. Volunteers for the Mir missions were few. The major exception to the common NASA attitude was Mike Foale, who behaved as should a rookie cosmonaut, and fully lived and worked with his crewmates. David Wolf was a close second and so too Shannon Lucid. However, these were personal decisions by the astronauts, not the result of any NASA system of integration. NASA as an organization by 1998 was little closer to understanding the Russians-only far more tired.

The other major problem was international politics, as the climate for cooperation evaporated within a few short years. In October, 1998 Boeing was fined $10 million for having held technical meetings with Energia and Yuzhnoye regarding Sea Launch without the presence of State Department officials. Perhaps even worse, the State Department laid down a stipulation that the countdown for all Sea Launch vehicles must be done in English and an American had to press the launch button. The Russians and Ukrainians were incredulous and furious. Counter suggestions included counting down in both Russian and English and having a Russian press the launch button, echoed by an American out of view. Even the Norwegian partner joined several times in both written and verbal condemnations of the American behavior while overseeing the commercial use of this Russian-Ukrainian launch vehicle.

The real problem was Loral Aerospace. On February 15th, 1996, China's Long-March 3B exploded as it lofted a Loral satellite built for the international satellite consortium Intelsat. Loral was found guilty of providing sensitive information to the Chinese which allowed a correction to the design flaw. The aerospace company was later slapped with a $20 million fine. Compounding the issue was that Bernard Schwartz had become one of the largest private donors to the Clinton Administration, which transformed an industry incident into a political scandal. The Clinton Administration's transfer of oversight of satellite export from the State Department to Commerce, which Schwartz had championed, was now reversed by a drumbeat from the Republicans in Congress.

In June of 1998, the return to earth of Andrew Thomas marked the end of Phase One. Tensions on both sides were strained and key officials exhausted. The Russians were angry at being implicitly accused of wrong doing in the Sea Launch matter. It was, after all, their rocket and their technology. Even more anger was directed at the aerospace contractors who blatantly boasted of the transfer of technology from the Russians to the Americans from Proton or Sea Launch or space station cooperation. "What should we do, rhetorically demanded Derechin, "go sue American contractors in American courts?" In response, the Duma imposed restrictions on sending technical material to NASA, even for previously agreed upon documents. That became a huge headache for Legostaev.

On the American side there was the frustration that the Russians were late with the delivery of FGB for the International Space Station, concerns that the Russians were stealing our technology and a belief that some in the Russian government were stealing American funds intended for ISS development. NASA officials were also fed-up with the perceived Russian stubbornness in negotiations. Every little request from NASA was fought over. Frank Culbertson sought to balance the demands from within NASA, Congress, industry and the White House in dealing with the Russians. The former astronaut and Navy commander was one of the few NASA officials who tried to see both sides. He gave a talk in 1996 in which he told of his wife asking just why the Russians walked everywhere in Houston, and why they brought their own food and why they dressed so funny. Culbertson told the audience that his wife then asked what it was like living in Moscow. "Well," deadpanned Culbertson, "we walk everywhere, we bring some of our favorite foods and I guess we dress funny."

Humor aside, never has a federal agency been less prepared than NASA was towards working with Russia. But it was not NASA's sole fault. It is a space agency, not a think tank for international business experts. The White House dumped a major diplomatic requirement on a very willing Dan Goldin and unwilling bureaucracy and then turned away.

In our Washington office the going was equally difficult. Congressional support for cooperation was dwindling, and the financial markets were leery to get involved, given the changing political mood. All these strains trickled down, far, far down, impacting on even the box sitting underneath the petrified salami. Inside the box were electronic chips manufactured by an American electronic firm called Data Device Corporation, (DDC). These parts were highly prized components for space hardware. Specifically, the parts in the refrigerator were designated as BU-61582, with the formal title of "Space Level Mil-Std-1553 Advanced Communication Engine Space Terminals."

The engineers at NASA's Johnson Space Center introduced the DDC parts to the engineers at RSC Energia. The space agency elected to use these components as the means to have each space station module communicate with each other. As I understood, it meant that the European module, the Japanese module, the American modules and the Russian modules all needed the same electronic 'brain' or switching system. These parts were that switching system.

Officials at Johnson Space Center arranged for Energia's EEE department, led by Vladimir Branets, the respected gray-haired veteran of space communications and electronics, to receive the DDC components. They were apparently shipped from DDC's factory to Johnson in Houston, and then directly on to Energia in Korolev, Russia. Once at Energia they were put to use in the two major Russian built modules for the International Space Station, the FGB and the Service Module. Branets was so impressed by the capabilities of the components and others from America that he persuaded Semenov to authorize use of American designed components in their Yamal series of satellites, in Sea Launch, the Soyuz rocket, and even in the Mir space station.

Branets told me it was a tough blow for many of the General Designers, including Semenov, to accept, but over time he agreed to de-emphasize Russian electronics in favor of the American and European. "Someday we will be able to compete again," patriotically promised Branets, "but now is not the time." And so at the next General Designer Council meeting in early 1996, the decision was taken to use the best electronic parts available, no matter the country of origin. Soon after that meeting in Russia, our Energia Ltd. office was asked to handle a new task, the ordering of those U.S. components requested by Branets for the commercial programs. The ordering and export of parts destined for use on ISS was done through NASA. We would handle the rest. Faranetta threw himself into this task, interfacing with DDC and the other parts manufacturers.

The program grew very nicely from 1996 until early 1998. We worked with the Commerce Department on eleven occasions and received the necessary licenses. State Department approval was required for a far more sophisticated part, but again the export license was approved without problems. Electronics for space deemed more sophisticated and more likely to be used in a military system had to be approved by the State Department. But we anticipated few problems, as the export destination was RSC Energia, a Russian company known intimately by the American government, and, after all, Energia was using the American parts to better integrate into international space programs. I felt isolated from the growing spate over technology transfer, though Faranetta took it far harder than me. He worried that our office could become a convenient target for some politician looking to score political points.

In late 1998 emotions reached a fever pitch when the Chinese were caught stealing nuclear secrets with the help of an American researcher at Los Alamos named Wen Ho Lee. He was arrested, though later the researcher was released with an apology. Meanwhile, the State Department and Congress ratcheted up pressure by accusing Russia of supplying Iran with missile technology. Boris Yeltsin had given Yuri Koptev a senior role in doing business with Iran and hence the Russian Space Agency, now known as Rosaviakosmos came under Congressional attack. A resolution was passed prohibiting the transfer of further Congressional funds to the space agency. The name change was indicative of the problem. Instead of being focused only on space, Koptev was now head of an agency tasked with both aviation and space issues. And aviation could include missiles. Indeed, I was told that the business of aviation now occupied far more of his time than space.

In the face of this confused situation the State Department froze all technology exports to Russia. We had come full circle, with the same situation as ten years before, when Payload Systems was seeking its export license for the black box to the space station Mir. Except now our two countries were partners on a multiyear, multibillion dollar space project.

Where was the leadership to push aside the ISS as an exception? The answer is that the Washington political community grew even more distrustful of the Russian motivations, still confusing the government actions (or inactions) with that of industry. One example will unfortunately suffice. In the midst of all these problems the long delayed launch of the FGB was finally nearing. A Congressional space authority sought me out to bet an expensive dinner that the Russians would purposefully destroy the Proton carrying the first module of the International Space Station, in order to force America to pay more funds to Russia. Her offer took place in full view of a half-dozen colleagues. I took the bet, feeling sick. If a leading Congressional adviser believed the Russians would purposefully destroy the first module, what chance was there for understanding? The launch thankfully went off without a hitch, but I never got my dinner.

If not for the close working relations between NASA and Energia, the prognosis for our export license would certainly have been bleak, as those DDC components were also the electronic brain of the latest fighter jets. The problem can be traced to that phrase "Mil-Std" which stands for 'military standard.' Those little components have been radiation hardened. In the event of a thermo-nuclear war they would be ok. Just like the petrified salami.

We waited and waited for approval. Nothing. I called into State and received no real answer. Chris and I discussed hiring consultants who specialize in assisting in tricky export issues, but Semenov, with his usual common sense, suggested that if the State Department did not wish to do business with the Russians, and wanted to freeze cooperation, that is a good thing to know, right?

After several months we finally received the rejection. Our request to send the parts to RSC Energia had been declined. No reason given. I immediately complained to John Schumacher. He turned me over to some bent-out-of-shape petty NASA official who called screaming that it "would be a cold day in hell when these parts go to Russia," and that "this whole issue is beyond your businessman's perspective." Meaning, I'm sure, that his butt was already on the line and the last thing he needed was someone with good reasons to export to Russia. In all the years of working with the Russians this was the rudest conversation with an American government official. All the other times there was always a strong desire to keep it polite, and even try and suggest how we might together solve whatever problem was the subject of the call. Losing my temper, I explained to the petty official that the EEE parts had already been sent to Russia by NASA. The application denied by State Department was a request to use them for commercial purposes in other Russian space programs. He didn't believe me and there the phone conversation abruptly ended.

Two days later he let me know that his office had reviewed my State Department application and as I had written (my signature was on the application, not Faranetta's) that the Russians had already been given the parts by NASA, I was, he informed me sweetly, open to being prosecuted, as lying on an export application was a federal offense. This was getting awfully personal, awfully fast. I had Branet's department quickly send documentation showing they had indeed received the components from NASA. I faxed the proof over to the State Department and NASA, demanding the rejection be reconsidered.

There were several consequences of this back and forth. The first was that Chris became increasingly agitated that we were heading into a dangerous situation. He decided to call it quits. It happened pretty fast, with him just one afternoon letting us know that he wanted out. We worked an agreement that stipulated that from that day forward he was no longer responsible for any actions taken by Energia Ltd. His leaving so quickly hit me pretty hard. Derechin was also very upset by Faranetta's decision, and so too Semenov. It also left me feeling exposed, and worried that I was staying with the ship despite the political warning signs. Fortunately, nothing ever happened to our office, but it was a pretty tense time, especially with Derechin and Legostaev seeking to understand just why Faranetta had left.

The other impact of the incident was that the Johnson Space Center engineers had the heat turned up on them. Rather than State Department admitting that it made sense to send the EEE parts to Energia, the export officers turned their wrath onto the NASA engineers. I have a hunch they cut some corners to send the electronic components to Moscow and thought that would be the end of the story. I flew down to Houston to discuss better coordination in the future. It was a useless trip. The engineers sat there too frightened to speak. I was never left alone. One of the poor shaking engineers followed me into the bathroom, literally, just in case I had memorized some vital layout at Johnson and would then write it down on toilet paper and flush it directly to Korolev, Russia.

Over the next two months I continued to argue with both NASA and the State Department. Semenov and Legostaev were calm. For these veterans of political situations, understanding the rules was priority number one. Given that the signals from the American government were clear, they were fine. Derechin was different, lashing out at the illogical behavior of State. So too those at Rosaviakosmos and the Duma. The ban was seen as an insult to the cooperation that had painstakingly been developed.

Sadly, most of the held-up parts were intended for commercial projects like the Yamal satellite program. This was a pet program of Semenov's, a new satellite venture intended to bring modern communications to the former Soviet Union. Don't we want that? Semenov told me there were literally hundreds and hundreds of forgotten regions in Russia that were no more than company towns of the gas company Gasprom. Energia had created Gascom, a joint venture with Gasprom to bring a sports channel, several news channels and telephone capacity into the farthest reaches of Russia. These were the sorts of towns and regions that were still voting for the Communist Party, and had been left behind by the economic and social advances taking place in St. Petersburg and Moscow. Semenov described weekly visits by Gasprom's huge helicopters, which would ferry in the newspapers, fresh food and supplies for these company towns. Creating modern communications for these villages is good, right?

But the most powerful argument involved Astronaut Bill Shepherd.

Shepherd, or Shep as he liked to be called, was scheduled to fly to the International Space Station in a Soyuz in the fall of 2000. This would be the first permanent crew to NASA's Station. Shepherd was to be the commander for crewmates Sergei Krikalev and Yuri Gidzenko. Don't ask why the American Shepherd was the commander of this Russian crew flying on a Russian rocket to the Russian built module. It was yet another little offense that went unnoticed by our press and our industry but reverberated through Russia. One cosmonaut quit rather than be part of this political charade.

Energia wanted to use the DDC parts to upgrade the Soyuz. That means make it safer for NASA astronaut William "Shep" Shepherd so there is less chance he would die in a failed launch. The State Department, in other words, was refusing to let Energia receive components to upgrade a Soyuz so it would be safer for a crew that included an American. The State Department refusal came despite the fact Energia had previously received these same components for inclusion into their modules for ISS. I told the State Department in a phone conversation that I was going to tell Mrs. Shepherd that the United States Government was blocking efforts by the Russians to assure the utmost safety for her husband's launch. The State Department guy went silent. I further told him if anything God Forbid happened to that launch I would release my notes to the national press. He began to whine, saying there was nothing he could do, his hands were tied, it was a political decision made over his pay grade, and so on and so on.

To this day I don't know if there was a Mrs. Shepherd.

And the electronic components remained in the refrigerator.

The real issue has to do with the capability of the United States to embark on a mission and stay the course. It has to do with keeping politics out of the space program. Having made the decision to invite the Russian government to participate as a full partner in the International Space Station, only an egregious occurrence warrants any change. The Russian space community felt whiplashed by the changing American attitudes. Even Koptev felt this way. In January of 1998 the fifteen senior governmental officials representing the member states of the International Space Station finally signed the operating agreements. In an elaborate ceremony held at the State Department, Acting Secretary of State Strobe Talbott signed for the United States. Each space agency official also signed a commemorative poster, with Dan Goldin's signature prominently featured in the center. Back at the office Yuri Koptev handed the poster to me. "Here, take it," he wearily said. "I don't want to be reminded about these negotiations again."

The Russian and NASA space communities struggled on, trapped together. Energia continued the never-ending battle to have Russian politicians honor the promises of funding for the ISS. Brian Dailey finally realized the errors in the model established by him and Dan Goldin. "In retrospect," he observed in the NASA Oral History project, "one of the things that maybe should have been adjusted...was how we should have distributed the money. Rather than send the money through the Russian Space Agency, it would have been better to have sent the money through the factories themselves...because the Russian Space Agency and more specifically the government,...was taking the money."

When the State Department first refused our export license for the DDC parts Energia assumed it would eventually all get sorted out. It did not. Finally I wrote a memo to Semenov apologizing how "to my regret I must admit defeat." It was my biggest failure on behalf of Energia, the one time an honorable request could not be fulfilled, and to make matters worse, it cost us a lot of lost money.

Of small comfort was the philosophical answer I knew would be waiting for me from Victor Legostaev. Early in our relationship I had made a mistake and dreaded telling this powerful Russian space official that I had goofed. Legostaev leaned back in his chair upon hearing my report and lifted his eyes to the ceiling. Then he looked at me with what could be considered a small grin. "Do not worry," he sincerely counseled, "It is not your last mistake." That became our motto over the next years and through all the battles. It was the wry humor of a man who had seen so much and hey, it will only get worse.

I went to Moscow to discuss next steps. Leading me into the private office, Semenov motioned for me to sit down. Wasting no time, he asked a simple question. "If you are not allowed to export to Russia these parts for a Soyuz scheduled for ISS and carrying a NASA astronaut, how can we consider ourselves full partners in ISS? Ask your NASA friends this question."

It was a good question. Still is.

Beneath the petrified salami lies the evidence.

At the end of 1998 a NASA friend introduced me to a start-up space company struggling to come to market. I decided it was time to try something new; something that had nothing to do with international politics and Russia. Energia well understood the reasons for my pulling away, but requested the Washington office be kept open. Not only had Semenov approved the office, but believe it or not, Russian government officials, up to and including the Prime Minister, had apparently 'blessed' the office. It was an authorized company for conducting Russian-American space business. "It would take us years to receive a new authorization, if ever," Derechin wearily explained. So we agreed to conduct lower level business, including handling travel for Russian specialists, after I shifted my focus to the new job.

First, it was necessary for me to request a leave of absence from my boss Dr. Semenov. Seriously. Derechin counseled that it was the proper behavior on my part. The General Director listened with regret, but was comforted, he said, that the office remained. This was the man who kept a flight-ready Buran in storage at Baikonur, waiting for the economic conditions to warrant a resumption of that program. He understood waiting.

As we spoke that day, I had no idea if I would ever work with Energia and Russia again. I couldn't imagine a project enticing enough to cause me to jump back into the struggle with Semenov against NASA.

It was of some comfort to realize how much had been achieved since 1991. Energia was now a private company with international shareholders. NASA had been forced to reform and pay for some of the Russian hardware and services. The industry had begun to realize how more robust and safer was our exploration because of having Russia's participation. Yet America still clung to the Cold War model of a huge centralized space agency that pushed aside innovative commercial projects. The next battlefield for industry reform would more than likely be within the halls of Congress. At least, that's what I thought.

Of course there was a document that needed to be signed and stamped before I could leave. In front of me Semenov signed the document, which granted permission for Energia worker J. Manber to take a leave of absence while maintaining the overseas office. No doubt somewhere, in some filing cabinet, that authorization still exists.

Section Three: The Launch of MirCorp 1999-2000

It is extremely unlikely that MirCorp will be commercially successful. MirCorp president Jeffrey Manber has stated that his company will offer Mir for advertising purposes, space tourism, and private scientific and pharmaceutical research. In addition, he has said that Mir could be used to broadcast live images of the Earth and to repair or manufacture space satellites. There are severe hurdles to generate significant revenue from any of these endeavors and some of them are so unrealistic that they call into question the credibility of the entire MirCorp effort.

- Dwayne A. Day, Florida Today On Line

Chapter 15: Rick's Audacious Plan

By September of 1999 the Mir space station was unmanned. The Russian government lacked the funds to pay for the construction of the supporting cargo ships necessary to ferry supplies to the station, and also reboost the station, which was drifting ever closer to the earth's upper atmosphere. The Russian government also had ongoing construction and operation obligations for the International Space Station. During this month the Mir's lowest point, known as the perigee, was now just 365 km above the earth. The Russian Space Agency had announced that the Mir would be forced into a controlled reentry in early 2000.

One day in early September, 1999. Rick Tumlinson was calling. "I got the man interested," he whispered in his best conspiratorial voice.

I took the bait.

"Rick, explain." After all, Tumlinson was the 45 year old pony-tailed, fast-tasking, scheming, manipulating, clever Rasputin for mega-wealthy Walt Anderson's space activities.

In recent months Rick had provided the philosophical basis to Walt for the effort to keep the Mir space station in orbit. It was Rick who launched the "Save the Mir" campaign, which until now had consisted of press releases deriding NASA for its pressure to kill a perfectly good station. Tumlinson had also fired off letters to Yuri Koptev. As a favor to Rick I had delivered the letters to the Russians. But other than that, I stayed far away from the "Save the Mir" campaign.

Not that I was neutral on the Mir coming down, of course. It just seemed that time had finally run out and that nothing could stop the momentum of the Mir's death. I had recently spoken with Alexander Derechin and Victor Legostaev, and both had said the Kremlin was wrapped up in joining the NASA space station for a host of political reasons. A senior Kremlin official who I had met at one of the Gore-Chernomyrdin receptions confirmed that it had little to do with funding and a lot to do with pressure being applied by our State Department. "Your government wants to tie the hands of the New Russia," he explained over a fine bottle of French wine. I wondered if it would work.

My friend just shrugged. "We ourselves have tried since Petre the

First," he paused to sip the wine, "to join with Western nations. It will work until it no longer is working." Fine. I just figured that the best course of action was to let the Mir die a natural death. If Russia couldn't muster the political will to save it, what the hell could I do?

Rick continued to whisper into my ear. "Walt, man, Walt Anderson. He's ready to buy the Mir and save it."

"Rick, you can't buy the Mir."

"Aw, c'mon Jeff, you just love those Russians too much. You can buy anythinnng in Russiaaaaa." And here he stretched out the words as to allow our collective petty imaginations to include whatever our little brains wanted to include.

Rick worked that way. At one moment brilliant in his analysis of a situation in the space community and the next veering off into some off-color impolite innuendo's.

"Rick, let me explain. The Mir is a national resource and it is not for sale. These people aren't the Brits, they won't be selling their national assets to foreigners just because times are tough." I reconsidered. "O.K., you think you want to buy the Mir? I can do that. I can get you a piece of paper signed by a government official giving you the station."

Eagerly he took the bait. "Gimme a ballpark figure."

"Because we're friends, $20,000." There was silence on the phone. I could tell that he was suspicious at the number selected at random. "Rick, if you're serious and listen closely we can perhaps enter into an agreement that reflects their sensibilities and answers their political concerns and gets you what you want. But you also gotta understand: I've seen the documents; they are planning to bring it down, maybe by late December, maybe in January. The Mir is coming down."

Rick was crestfallen for maybe a day. Maybe two days. He behaved that way, disappearing for no reason at all, as if he had some sort of second life somewhere in Los Angeles. Then he resurfaced filled again with optimism and energy. "I've been speaking with the man about you." In that same conspiratorial whisper.

"So?"

"He wants you to arrange everything. You're at the center of it all. You're the only one who can bring the Russians on board. They trust you, man. We just need them right now and then we can run the Mir with western management."

"No. Sorry. No can do. Not interested. The Mir is Russian, Rick. It will be that way forever."

"We got the money and we also got the technology. Don't fuck me on this one, Manber."

"What technology?"

"Don't forget about the tether, man, you helped us with that, or did you forget?"

Oh, lord, there's reason to this man's madness.

Hanging from a Heavenly Thread

Tumlinson had obviously been giving this some thought. Long known in the space industry for his emotional diatribes against NASA, he had recently, with Anderson's financing, turned less vocal and channeled some of that frustration into funding meaningful technological projects and commercial space ideas.

My favorite was a revolutionary project that involved deployment of a tether in outer space. Rick was funding this effort through his Foundation for the International Non-Governmental Development of Space (FINDS) and had worked with Chris Faranetta to have Energia study its technical feasibility.

Funded with an initial donation of telecommunication stock from Anderson, Rick was now sitting on an endowment worth over $15 million. His mandate, under the careful eyes of Anderson and Anderson's friend Gus Gardellini, was to scatter investment funds across a range of commercial space projects. After years of being ignored by major space policy officials, the dreamers had invaded and Rick was genuinely influential. With a little bit of luck and good investments, Tumlinson through FINDS could change the balance of power away from NASA and towards more commercial projects. NASA usually focused their funding on hardware projects that buttress their own goals, and not projects that were the most promising. NASA officials then would testify before Congress backed by dozens of contractors and subcontractors that supported their latest boondoggle, whether space shuttle, or Freedom, or the International Space Station (ISS) or some other overpriced hardware scheme. But perhaps FINDS could change that. Imaginative researchers long shut out by NASA had ideas that were low-cost and practical. Like the tether.

The tether is a product of beauty; modern art slicing through the heavens. It is a long thin sliver of wire deployed in orbit. Long a dream of space planners, it seems possible to capture energy that is drawn down the wire, like using Ben Franklin's kite in a thunderstorm to gather electricity. In March of 1994, an engineer named Joe Carroll deployed a tether from an unmanned Delta rocket that stretched some 19 kilometers out into space for about four days. It was long enough and reflected enough of the sun to be seen easily by the naked eye. No one had ever deployed a tether that far out and for that long.

The Italians had tried from the space shuttle and failed. The wire refused to spool out and remained snagged. The Russians were interested in undertaking a similar experiment. Tumlinson proposed such a project, bringing together the Mir with Joe Carroll. The practicality of the project was obvious. If deployed successfully, the tether might draw in enough energy to reduce the number of Progress fuel ships required to fly to the Mir or any future space station. Some combination of the two, a working tether and money, might now do the trick in keeping the space station alive.

During the summer of 1999 FINDS commissioned RSC Energia to study the tether concept for a $100,000 contract. The study, led by Vladimir Syromiatnikov, concluded that it was quite feasible. Everyone, from Semenov on down, was excited about the possibilities.

The basic ingredients might be coming together to save the space station. There was a technology path, a funding path and an overarching philosophy.

There were two problems.

Time was one. The Mir was falling out of its safe orbit. It had been unmanned for months and within weeks a Progress rocket would blast off on a mission to link to the doomed station and push it to its fiery death.

The second problem was political. Rosaviakosmos had already agreed to bring the Russian space station crashing back to Earth. The Americans wanted the station down, so Semenov would focus all of his attention and limited resources on the International Space Station. Even though I was no longer involved day to day with the Russians, it was clear that NASA had handled the situation involving the end of Mir with its usual lack of understanding. Goldin simply ignored Semenov and pushed through a deorbit agreement with Yuri Koptev. The space agency either ignored or was in the dark regarding the Russian government decree that had turned the Mir over legally to RSC Energia.

I had heard Semenov was mad as hell regarding the heavy-handed treatment by NASA. So maybe he would support outside funding, but would never allow foreigners to take control of his station. Tumlinson would hear none of this. "Get off your butt man and make some history!" Reluctantly, yet also intrigued, I agreed to meet with Walt. FINDS and the Space Frontier Foundation were holding their annual convention right by the Los Angeles airport and Walt and his advisors, as well as Rick and the designer of the tether, Jim Carroll, would all be attending.

Why not sit down with Anderson? After all, the Mir had long ago gotten into my head. It seemed to me worth the flight to Los Angeles to hear out Rick and try one last time to save that wonderful "fixer upper," as astronaut David Wolf had referred to the Mir.

Chapter 16: If You Can't Buy, Then Lease

It had been some ten years since I had last seen Walt Anderson, not since heaping abuse on a project he was funding. The project was for a new commercial launch vehicle developed by Chris Roberts, a friend and entrepreneur, along with Bruce Kraselsky from the Commerce Department, and Peter Diamandis, who later created the X-Prize.

The business plan for the new launch vehicle seemed doable until the section that read something like "given our management philosophy, we will not be accepting any contracts from organizations associated with the Pentagon, or any military organization whatsoever." Maybe my reaction was overboard, but I stopped reading and wrote in large block letters. "This is bull, a waste of time. This view alone will prevent work with your only sure customers, NASA, and DoD." Then I sent it back to the potential investor who had requested my opinion. Walt was furious. At a party a year later, he walked right past me several times, not saying a word. Walt was

Telecommunications entrepreneur Walt Anderson invested in dozens of private space projects. He believed passionately in a space exploration effort without the iron grip of NASA, and backed his belief with tens of millions of investment capital.

Photo: Gus Gardellini

passionate in his hatred of the military. An honorable view perhaps, but not practical if you are investing in a market where the government was the largest customer, as was the case in the early 1990's for the small launch vehicle industry. So there was some nervousness about meeting with Walt.

Spotting each other in the lobby of the Marriott just across from the Los Angeles airport, about thirty minutes before we were scheduled to meet, Walt was classy enough to quickly comment that my criticisms of his rocket project had been correct. That put me a bit at ease.

Anderson looked about the same as the last time we had been together, but the hair was prematurely white now. At 46, he still had a youthful slim build and dressed casually, almost too casually, like a well-to-do software developer. Never in a business suit, never in a tie, he was the very model of the American businessman of the new century. His skin was pale and his manner awkward, like a teenager that hung around the computer too much.

Tumlinson was nervous about the meeting. On the phone he had injected without warning that Anderson was somewhat like a recluse. Like a Howard Hughes type of guy.

"Like Howard Hughes, Rick? With long fingernails and fear of germs?"

Tumlinson was embarrassed as I pushed him about Anderson, but he didn't back down and even in the hotel, while having something to eat, he was warning about how Anderson behaved and what he wanted and didn't want. It seemed that Anderson was shy and avoiding meetings outside a small circle of his advisors. Anderson wasn't, promised Tumlinson, a kook. Or was he? I mean, Tumlinson was promising that Anderson wanted to buy the Mir space station. That sounded like something Howard Hughes would have done. I chalked up Tumlinson's concerns to his need to micromanage any situation.

Since I had last seen Anderson, his investments had apparently skyrocketed. By all accounts he had earned several hundred millions of dollars in telecommunication and Internet ventures. He had founded a company called Mid Atlantic Telecom, a regional long distance carrier, in 1984, to take advantage of the break up of AT&T. Anderson also founded a European telecommunications company called Esprit Telecom, which went public in 1997, and from which he got a good chunk of his money. Now he wanted to pour his wealth into creating commercial space ventures. Most serious space ventures at that time had Anderson's money helping them in some form or another. He was the lead investor in a Los Angeles project

IF YOU CAN'T BUY, THEN LEASE

called Rotary Rocket, among who knows how many others. Without a doubt, much of the more creative American space entrepreneurial ventures of the time were supported in one way or another by Anderson.

Walt skipped all preliminaries after again apologizing for dissing me ten years before. He immediately asked what I thought of buying the space station. Sigh.

"You can't buy the Mir."

"Why the hell not?" His eyes got wide, and his jaw jutted out which is how Walt took in unexpected and unwelcome information. "I hear you can buy anything in Russia."

I was not looking forward to a long educational program for a bunch of space dreamers, no matter how wonderful the goal. But I tried to be nice. So I carefully explained again the political situation to Rick, Gus and Walt while huddled in a dark corner of the lobby. How the Russian space community had not fallen as hard or far as the rest of the society. How the space station was very important to their psyche and better to ditch in the ocean then sell out to foreigners. How Energia was a private company that had been given control of the space station. And finally, how Semenov tried to introduce market forces both within Energia and his subcontractors and across the parts chain, up to and including NASA. I figured that would be important to a guy like Walt, who lived and breathed capitalism and free markets.

What was so brilliant about Walt is that the guy listened. He really listened, though his eyes roamed over the hotel lobby. Standing back at a discreet distance were a dozen space entrepreneurs waiting and hoping for a chance to talk with the man who was committed to bringing about the commercialization of space exploration. It was a strange disconnect-their projects were no doubt measured, taking a market niche here and there while NASA wasn't looking.

Yet here we were speaking about buying the goddamn Russian space station. What more could one do to antagonize NASA? Keeping the Russian space station alive would piss NASA off, and threaten the reputation of Dan Goldin more than any other single commercial option. Walt's questions were rapid fire. Very direct.

"Is Energia corrupt?"

"Do you have to pay them money under the table? I won't do business that way."

"What do they need in terms of money and when."

"Can we get control or will they make us compete against themselves for Mir services."

"How much do we need to get going? Twenty million, thirty million dollars?"

He never agreed or disagreed with me. He asked and I answered.

After a dozen or so questions we went into the scheduled meeting in one of those horrible windowless hotel conference rooms. Waiting in the room was Vladimir Syromiatnikov. Never at a loss for words, Vladimir regaled us with his work on Sputnik. He was responsible, he explained, for the one moving part on the Sputnik: the fan. That's hard to top, listening to an engineer whose personal experience in space began with humanity's first satellite. Walt lost patience at Syromiatnikov's tale and grabbed control.

"This is the first serious meeting to explore whether we can stop the Russians, aided by NASA, from destroying Mir," he solemnly began. He rattled off many of the points I had raised with both him and Rick. We won't be going to Russia to buy the Mir, it belongs to the Russian people," explained Anderson. "Who the hell needs the problems of ownership? We want to work with them. We don't want to embarrass them. We need to work something better out."

Fortunately, Syromiatnikov agreed emphatically with Walt's talking points. Listening to Walt I felt a lot of respect. He had absorbed the points and was testing them on his group of advisors. These were now the basis of his starting position.

Then he and his lawyer, Jim Dunstan, an earnest man with a baby face and a bright smile, who has never outgrown his love of space, began discussing just how to proceed. I think it was Jim who suggested they offer Energia a commercial lease. "If we can't buy it," mused Dunstan, "then let's offer to lease it like a building."

Walt's eyes lit up. "Say," he said brightly, "that's great. What do you guys think?" He looked over to Vladimir and myself. We agreed, though both carefully stressed over and over that it was hard to predict what Semenov would think of the plan. It is a point difficult for Westerners to accept: that in a company of some 18,000 workers the decisions were still made by one man and one man only: Yuri Pavlovich Semenov.

Before the meeting broke up I wanted to hear from Anderson an estimate of just how much money he was prepared to spend. Rick was reluctant to voice a number. "Hey, remember, Manber works with the Russians!" For me it was a simple question. If it wasn't enough, let's not waste our time. Walt understood. "If," and he stressed the 'if', all conditions are met. If the cash flow is transparent, if they work with us and not against us, if there is no Mafia and no problems, "then I'll put aside $20 million to begin the project."

Twenty million dollars. A huge amount in a normal industry. A good-sized amount in space exploration and enough to get Semenov to take these guys seriously.

For the first time I had a funny feeling. Maybe this was for real.

The meeting was over in an hour. In back to back meetings it was decided that Anderson was no longer trying to buy the Mir; he would want to lease it. I would discuss the proposal with Energia's senior management and Syromiatnikov could echo some of my analysis of the current situation.

Then it was time to see some Los Angeles friends, to the surprise of Walt. But I've never really been comfortable in industry get-togethers, even the high energy of the Space Frontier Foundation meetings. What's known as the New Space community has so many ideas and few paths for implementation that it can be frustrating. Leaving the hotel lobby it was easy to spot Anderson, the ultimate angel investor, holding court over a crowd of New Space entrepreneurs.

Chapter 17: The Merchants of Space

Just two weeks had passed from Rick's phone call in early September to the Los Angeles meeting. From the first phone call from Rick I had kept in touch with Alexander Derechin.

At first very skeptical, given the short period of time left to the Mir, he was soon worn down by my daily calls, and my background report on Anderson. Helpful without a doubt was the fact that Tumlinson and Anderson had funded the tether project, so they were a known group to the skeptics at Energia. Still, it seemed too audacious for Derechin to take seriously. Finally, I wrote a fax that said simply: "Mr. Anderson is prepared to spend more money in the next six months with Energia than NASA and Boeing combined." That first got Derechin's attention, then Legostaev and finally Semenov. The space veteran agreed to meet with the American entrepreneur.

Fueled by Walt Anderson's financial support, Rick Tumlinson and his FINDS organization was soon being welcomed by the likes of Ray Bradbury and other visionaries. All sorts of creative and crazed space entrepreneurs gravitated to Anderson and Tumlinson seeking support. Photo: Gus Gardellini/Rick Tumlinson

That was a meeting I would enjoy. The veteran Russian space official who had the guts to pursue commercial space exploration meeting with the younger, wealthy iconic example of American jungle capitalism. Both found themselves fighting against their own governments. Both saw nothing wrong with making a buck from space.

But first there would be an intermediate step. It was agreed that I would introduce Rick Tumlinson and Gus Gardellini to Semenov. Anderson didn't want to go to Russia unless Tumlinson felt it was worth pursuing. Also at the meeting would be a Los Angeles consultant named David Anderman. I didn't know him well, but like Chris Faranetta, he had soaked up all sorts of data on Russian hardware systems despite a lack of technical education. I had heard David had been a musician in Los Angeles, before jumping into the space business. Most recently, he had taught himself to read Russian, and it was Anderman who had first raised the idea of buying the space station Mir. During a board meeting of the Space Frontier Foundation back in the Fall of 1997 Anderman had proposed both to start a "Save the Mir" campaign and also that an effort be made to capture, in some way, the space station itself. With encouragement from Anderman, Rick ran with the "Save the Mir" effort.

An invitation from Energia was hastily pulled together, the visas were arranged, with Rick grumbling the whole time about needing a passport, flying overseas, not really wanting to visit Moscow. The trip didn't bode well for the project.

First, Semenov didn't even want to meet with these young men who had no serious business with Energia. Derechin received approval for the meeting only at the last minute, after we had arrived. The whole time Semenov sat there glumly looking at Tumlinson spin his tale of saving the Mir, and screwing NASA, and giving it to those NASA Congressional loyalists. Semenov sat in his usual seat watching this man with the long ponytail tucked in the back, and obligatory baseball cap. For this occasion, Rick wore into the meeting a green "Save the Mir" cap which he placed in front of him on the conference table. The legendary conference table around which the great Russian pioneers of space had sat. And now there in front of Semenov was a green baseball cap. Gus wore no tie and a buttoned up black shirt, fashionable perhaps in Miami or Los Angeles but not quite in Semenov's Moscow. Their dress, their emotional message and the fact that they were not the source of the money made the meeting a waste of time from Semenov's perspective.

It reminded me of a meeting we had once had in Washington. Semenov and Legostaev had been drinking all evening with George Abbey. The Russians had agreed to meet with an American space entrepreneur who had a small launch vehicle company in Northern Virginia. He arrived at our office dressed head to toe in black spandex, having cycled to the meeting. The American businessman eagerly explained that he was building a launch vehicle that would develop eventually into a manned rocket, and wanted Energia as a subcontractor for the project. The two senior space executives, veterans of dozens of manned flights, listened, thanked the young man and watched as he gathered up his helmet and went off on his bicycle. Professor Legostaev just stared at the door, as if still studying the spandex-entrepreneur. Then he issued one of my favorite comments. "If it was so easy to build and launch a manned spacecraft," remarked Legostaev, "other nations would have done it, no?" Someday a private company using private funds will develop a manned launch system. Maybe Internet entrepreneurs like Elon Musk or Jeff Bezos, now racing to develop commercial unmanned rockets. Not today.

After maybe thirty minutes Semenov interrupted Tumlinson to explain about NASA's pressure on Russia to quickly bring the Mir down. Semenov was candid, speaking of the behind the scenes threats, the dangling offer of millions of NASA dollars if the Russians destroyed the Mir, the linkage being imposed by the Americans between the Mir situation and vital strategic discussions with State, meaning that the Russian defense establishment was being denied routine requests until the Mir was destroyed. Rick went ballistic, saying he was going to tell his media friends.

I had to jump in while his comment was being translated and remind Rick that the details of conversations in the conference room were always kept in the conference room. That just made Rick more nervous and clearly shaken. It was the wrong setting for Tumlinson. He was out of place in the stark setting of the room, the translators, and the older men sitting staring across the table-but in the end Rick achieved what was necessary. Semenov agreed that if conditions were right for Energia a partnership with Anderson and his team for the preservation of the Mir could be discussed. Semenov warned Tumlinson that time was very short, since a Progress was being readied to bring the space station down. At the same time he made clear his distaste for NASA and how it did business. That was an important statement that I knew Rick would take back to Anderson. With a heavy sigh, the General Designer stated his willingness to battle his own government if he had the funds.

It was not an easy step. That the tether project and the Anderson proposal was being done not by people from Boeing or Lockheed or Mitsubishi was upsetting to these space engineers, all of whom sought to work with their foreign aerospace colleagues and not with unknown entrepreneurs. But the chill in Russian-American relations was strangling that sort of cooperation, and perhaps the answer was this new group, a combination of deep pockets and crazy ideas.

"My whole life I have worked with old men," Semenov mused to me the next day. "Look where it has gotten us. Maybe the future is with younger men now."

Younger men to save the aging space station.

Capitalists to save a station built by Communists.

A telecommunications multi-millionaire to work with a General Designer appointed by Communist Party boss Mikhail Gorbachev. And all this to battle against NASA's continued avoidance of open markets.

Strange new world indeed.

Chapter 18: Risky Business

Mir's Perigee had dropped 14 kilometers in a month, to 342 kilometers in October. It was falling faster than the anticipated 1 kilometer per week due to increasing solar flare activity from the sun, which was entering the most active phase of the 11 year cycle for solar flare activity.

Walt Anderson was rich. That much was clear. But his wealth was not how he should be defined. Rather, he was a man in search of a holy grail, some magical elixir that would allow space exploration to flower and not suffocate in the government morass. He had made hundreds of millions in telecommunications and Internet ventures and thrown much of it away in romantic commercial space exploration projects. Few knew of the extent of the space investments. That was kept quiet. The wealth was clear. We planned to blast across the Atlantic in his Gulfstream jet, destined for

Rick Tumlinson (left) and John Jacobson playing Risk while flying to Moscow with Anderson, Gardellini and the author. It was a magical time for Anderson and his advisors. His wealth was increasing exponentially and realistically envisioned was the private development of an orbital infrastructure lead by a new generation of Internet and telecommunication investors lead by Anderson.

Photo: Gus Gardellini

Moscow's private Vnukova airport. Partly it was a gimmick to show the Russians that Walt, despite his non-affiliation with a known company, had the funds necessary to implement an agreement.

The meeting in Moscow on the "Save the Mir" project would be the last if it was not successful. There was just no longer any time to spare. Multi-millionaire Walt Anderson and General Director Yuri Semenov must meet and must learn to respect each other, and Anderson had to show he had the required funds.

In the month since Los Angeles Anderson had not stepped back from his dream to rescue the falling space station. He himself couldn't explain fully the reasons. I had asked. Both for myself and because Derechin kept asking.

Anderson was driven by a mix of cold-blooded capitalism tinged with the willingness to invest years and tens of millions of dollars in creating a climate that would allow for a free market in space exploration. That meant bringing down NASA, or at the very least destroying its hold on every aspect of manned space exploration. Pride, arrogance, intelligence, visionary, all these qualities were rolled up into Walt Anderson.

It was the sort of ingredients the men who ran Energia could respect, except for one. These were careful, very careful engineers, who took no risks that were not calculated. When Legostaev arrived in Washington we would review every step of the trip. Nothing was left to chance. The free market investor driven by passion ran counter to this attitude. Could common ground be found to work together?

Walt's gleaming white jet was parked at National Airport in Washington, DC.

What an extraordinary luxury for someone like myself who had made dozens of the long trips to Moscow from the States. My wife Dana drove me the fifteen minutes from our home to National Airport and watched as I climbed onboard.

Onboard were Walt and Gus Gardellini. Gus, 40, was the ever smiling reality check for Walt. Black-haired with handsome dark eyes, Gus was filled with energy and enjoying every minute of life. There was no eagerness on his part to give up time away from his apparent fairy tale existence in Miami, thanks to the millions made from investing in telecommunications projects with Walt. But he came since Anderson wanted his advice on whether or not to do business with Energia and with Russia. Also present was John Jacobson, a friend of Gus who had experience in Russian oil and gas deals. I was a little concerned about the inclusion of John; usually those Westerners with business experience in other Russian sectors arrive at

Energia with preconceived notions of doing business in Russia and end up insulting Energia, which by and by had remained a clean organization. I was wrong about John, whom I came to regard as a "gentle giant," always ready to help and far more astute than he let on.

Also present on the jet of course was Rick Tumlinson. A mess of his usual mixed up energy, Rick was peppering me with questions regarding the Russian's attitudes, whispering to me about the honor of being invited on the jet, constantly taking pictures for posterity and when we did finally arrive in Moscow, unable to find his passport.

As we took off to the north Walt was still on his phone. Rick was angering me by explaining this mission was a "slam-dunk," no problem. Energia would just fall into their arms in gratitude.

"Ever work with Russians, Rick?"

"Sure, I did the tether study project…"

I interrupted, trying to explain some of the Russian politics. That there were forces working against us that had already backed the Russians into a desperate situation with the Mir. Pockets of U.S. government officials had made it their goal to force the Russians to rid themselves of the Mir, so as to wrap their arms closer around the production line of launch vehicles and the factories that manufactured the missiles.

As the plane banked east over the Chesapeake Bay and soon enough the Atlantic Ocean, Tumlinson dismissed the political analysis and turned his frantic energy elsewhere.

Both Rick and Walt were naive regarding the strength of the political tensions. To them, it was a simple equation: the Mir was crashing down into the Earth because the Russians had no money. Given money, the Mir would remain in space. NASA would be forced to confront serious commercial competition for the first time in its existence. Confronted with competition, the space agency would change the blotted way it conducted its business. In a sense, therefore, Walt's motivation for spending tens of millions of dollars in Russia was not solely based on opening a new pathway to space but also as a way to force change within the U.S. space agency. A patriotic act from a man who would violently disagree that his behavior was at any time patriotic.

The cabin had large leather seats and a tastefully designed wood sound system console against one wall. The flight was expected to be smooth, so we all settled in for the long journey to the refueling stop in Scotland. I was ready to answer any question from Walt. On Energia. On Russia. On politics with NASA. On my background. On the dangers of Moscow.

"Well," Anderson asked to all of us, "who wants to play a game of Risk?"

Remember the game Risk? You take over the world by rolling the dice and moving plastic armies into the different countries? Remember when you were in university and had nothing else to do? Did I want to sit in a Gulfstream and play Risk for the whole flight to Russia, instead of sleeping or reading or preparing for the critical meeting?

I took the green colored armies.

Soon Gus and Walt and Rick and John and I were yelling and screaming and scheming for world domination. A pattern clearly emerged: Gus was ambitious but didn't care if he won or lost. He took over North America, which is dangerous, given one can be attacked from all sides, through Alaska, Latin America or the Northeast.

I was cautious, taking South America and Africa and trying to hold onto my geographical base until a strategy could be devised. John played cautiously as well. For Rick and Walt it was simple. These two guys were out to destroy one another through the countries of Europe and the vast emptiness of Russia and Asia. That was the plan of both men. Thirty minutes into the game we all asked about food, expecting some nice meal befitting a private jet. But it was not sushi, nor pasta. It was Dominoes pizza.

Well, the wine was good.

Gus was out of the game first. I followed fifteen minutes later. An agreed upon truce with Walt allowed me to solidify my positions in Africa, but caused Rick enough heartburn that he attacked me through his base in Australia regardless of the strategic merits. That seemed a point worth remembering.

I sat alone a few feet behind the table and watched them play, looking for some sign that Rick and Walt understood the seriousness of the situation. Or if they understood the obstacles that faced anyone trying to stop the de-orbit of the Russian space station, now just a month or two away. Though I had no official role in this effort I had already spent many long hours with Derechin walking through the problems that would have to be solved. Technical challenges. Political enemies to overcome. Contractual issues. Psychological hurdles regarding not only transfer of control of a Russian asset, but also the difficulties involved in our working together.

Time for some sleep. Let the maniacs battle between themselves.

Chapter 19: Time to Bottom Fish

Yuri Pavlovich Semenov was angry. It was a pent-up anger towards the gods of space for so conspiring against him and his plans to save Russia's space station. Everything was going wrong. There were no funds from his government. The Duma's promises had come to naught. His own space agency head, Yuri Koptev, was fighting to end the life of the Mir. His country had no money. Semenov had tried to find customers and investors for the space station and each promising lead had turned into a mockery of the proud history of Russian space exploration. The General Director was angry, angry at the world.

Sitting next to Semenov in the Energia conference room was Valery Ryumin, the burly veteran cosmonaut, and a crewmember on the final Shuttle-Mir flight. Ryumin was best known to the American space community for the controversy surrounding the final shuttle mission to Mir. NASA loudly criticized his participation during the final docking between

The first meeting between Walt Anderson and Yuri Semenov. The two men traded tirades and swapped horror stories on working with government. Anderson was the first true American entrepreneur Semenov had met and the Russian space veteran's desire to create a commercial space market appealed to the fiercely libertarian Anderson. Photo: Gardellini/Tumlinson

the NASA space shuttle and the Mir space station in June of 1998, as nothing more than a Russian junket. Here was a senior manager of Energia now being placed on the manifest as a shuttle crewmember. It was easy to poke fun of Ryumin, as many did at the Johnson Space Center. He drank too much; he smoked too much; he weighed too much for a cosmonaut; he never bothered to learn English for his shuttle mission. But there is a superhuman strength to the man, both physically and in spirit. Ryumin was a hero to many within the Russian space community. During a Salyut space station space walk in 1979 he had encountered severe problems and almost died while stranded during a spacewalk in open space, the sweat pouring down his visor making vision impossible and sharp pieces of the station looming all around. Finally, he was able to pull open the stuck hatch.

Unbeknownst to NASA, there was a secret reason for selecting Ryumin for the shuttle mission. The trusted advisor had gone to inspect the space station, and report to Dr. Semenov as to whether it was worth saving. As usual, the General Director insisted on first hand observation using only respected eyes. The veteran cosmonaut's conclusion was that saving the space station would be difficult, but possible. Had he concluded otherwise, Semenov would have abandoned the efforts.

There was intense political pressure brought by NASA to remove him from the STS-91 manifest. During this episode there came a reason for me to visit Ryumin in his Houston home. With him that evening was his daughter and wife Elena Kondakova, the beautiful cosmonaut who flew both to Mir and on the space shuttle, and today serves in the Russian Duma. Husband and wife together asked my advice regarding the U.S. pressure. I answered that it was a very sensitive time in relations between Russia and America and perhaps pulling back from the mission was the right political move. Anger flashed through both their eyes. I knew then there would be no retreat.

The importance of exploration for this space couple was again hammered into my head moments after Ryumin's successful launch on the space shuttle. We were celebrating in the Radisson Hotel down at Cape Canaveral. At the lobby bar were the top officials from Energia, including Legostaev, who had traveled from Moscow to show his respect for Ryumin. As the drinking was getting pretty serious Kondakova made her entrance amidst great yells and shouts from the Russians. Coming to the bar she stood next to me. After the congratulations in several languages on the safe launch had calmed down I also congratulated her on being free now.

"What do you mean, how am I free?" Her irritation clear.

I explained she was free because her husband was in space.

"No, Jeffrey, you have it wrong. He is the free one because he is in space now."

A year before it had been her turn to fly and Ryumin and I and two others had driven out the night before the STS-84 shuttle launch, so husband and wife could have a final pre-launch moment together. It was dusk and we drove silently near the deserted launch site where at about four in the coming morning the shuttle would blast off to dock with the Mir. Arriving at the low building that housed the crew, the rest of us stood back near the car, while Ryumin, big, burly, bearish Ryumin, stood outside on the lawn, looking up at the open window, while his wife leaned down from the second floor, like a princess from a fairy tale. Only this princess was preparing to leave the earth. Husband and wife spoke softly, tenderly, and then with final words of encouragement, these two fighters parted and she, within hours, blasted off into the frontier they both passionately loved.

Next to Semenov on his right was Professor Victor Pavlovich Legostaev. He had just returned that morning from the Pacific equator, out near a tiny speck called Christmas Island, where he had directed the second successful launch of the multi-national communications satellite project called Sea Launch. Legostaev had received a hero's welcome on his return, as the future of Energia seemed to be riding on every commercial launch. Yuri Koptev was refusing-or was unable- to fund Energia work on the International Space Station, leaving it to this commercial company to pay for what should have been federal commitments. Consequently, a single launch failure of Sea Launch could have put Energia out of business, and doom any chance of Russia meeting its international inter-governmental space obligations.

Next to Legostaev was Alexander Derechin, the head of International and Marketing. It had been a hectic month with both of us aware that there was a chance we would end up in a delicate situation of getting this deal done, but having to protect our own interests on opposite sides of the table. Derechin had taught me much about international negotiations, just as Hank Mittman from the Commerce Department had earlier taught me about contract language. Two good teachers.

On the far side of the table, furthest from us was the always silent Nikolai Ivanovich Zelenschikov. Americans never failed to comment how the number two man at Energia bore a startling resemblance to Abraham Lincoln, without the beard. He had the same craggy marked face with a large mole on his cheek and deep dark eyes. Zelenschikov usually said little-he was a man lacking the luxury of wishes or hopes. His was a world of precision. It was Zelenschikov's responsibility for each and every launch that involved Energia from Baikonur. The launch of the cargo ships and the safety of the crew was his operational responsibility.

Semenov sat, as Semenov always does, at the middle of the table, his back to the Birch trees right outside the massive windows. I wondered what Sergei Korolev, with his brooding face, who at the height of the Cold War fought alone against the American industrial space complex, would think of this last-ditch attempt to stop the ultimate insult: the bringing down of a flight worthy space station.

On my side sat Walt Anderson and his team. Walt was impatiently playing with his pen, a nervous habit when bored. He came all this way to talk business, not listen to tirades. Semenov was shouting about the predicament that he alone faced. President Yeltsin and the latest Prime Minister and the Deputy Prime Minister and the head of the Duma had all vanished, leaving Semenov alone to ponder how to solve this problem. The Russian politicians had fled for cover; understanding in that self-preservation inherent in all politicians that the re-entry of the Mir could be a political disaster. Russians had lost enough; and now their proud space heritage? And the international implications, should the massive space station not burn up but descend crashing into the earth, were too horrific to contemplate.

Oh yes, money was promised. Promises are so easy to make. Each and every year the Russian Duma patriotically and in timely fashion voted to provide funds for the Russian space station Mir; but in a page taken from Samuel Beckett's ironic play *Waiting for Godot*, Energia must wait for nothing to happen and nothing to arrive, for there was nothing coming. Nothing was all they had to wait for.

This was the source of his anger. Semenov was now emphatically pounding his fist on the glass-protected mahogany conference table that was so filled with Russian history. There was no worry about hurting this table; it must have been accustomed to quite a few emotional outbursts from the four men who have held the title of General Director. Today, in mid-October, there was only anger at these Americans sitting across from Semenov. The Americans audacious plan to save the worlds only space station was irritating to Semenov. Was this another waste of time? Another group of foreigners whose strings are really pulled by their government? Or self-promoters who will use the publicity to advance their own commercial business?

Despite the outburst, Semenov seemed in a good mood. When the General Director was truly angry, it was like being in the path of a tornado; it sucked the air out of your lungs and your focus was on self-survival. At the start of the meeting Semenov had looked over at me and bellowed his usual question: "Why aren't you sitting on our side?" It was his way of showing his trust in me in front of the visitors. Sometimes I did sit with Energia, especially when the visitors were not to my liking. This time I sat with Walt Anderson. That doesn't mean we were in for a pleasant experience. Rather than being greeted as saviors, Anderson and his colleagues were viewed as a waste of time by many across the table and across Energia's senior management. It was too late; the Russians reasoned. Not technically impossible, though the now-deserted Mir was sinking at a rate of several meters a day. In another month or so the massive structure will be dangerously low, skirting the upper atmosphere of the Earth.

There was another reason for hesitation. It was unspoken to these foreigners. But the head of Rosaviakosmos had publicly stated the space station is coming down. Should they cross Koptev for unknown foreigners?

Energia's Woes

Semenov in his tirade now laid out the technical situation. An unmanned cargo ship was being readied to launch into space. The plan was to attach this Progress cargo vehicle to the space station and push it into a controlled descent. The danger was considered high. No space structure this size had ever been built, and certainly never thrown back to the earth from which it had so brazenly escaped. Though the re-entry was considered less dangerous than Skylab, which on July 11th, 1979 came back uncontrolled as NASA, because of poor planning and the delay of the first space shuttle flight, lacked any launch vehicle to either boost the falling station or help control its return. Parts of Skylab did crash into the earth.

No, it was too late from a business viewpoint. There would be the inevitable weeks of negotiations, discussions, and understandings as to why these Americans wanted to rescue the aging station. What possible motive could capitalists have for saving this station? And if not driven by money, how could they have the funds required?

The Mir had been unmanned since the end of August, when Slovakian Ivan Bella and Viktor Afanasyev, along with French cosmonaut Jean-Pierre Haignere, completed a safe and uneventful six-month stay. The long duration was a gift to ESA, giving them an equal status in that department with NASA. They experienced none of the problems that afflicted the station during the NASA crews. No major problems with computers or oxygen generators. Despite the apparent health of the station, it was widely believed that this 27th crew was the final mission.

Dan Goldin cherished his so-called special relationship with his counterpart Yuri Koptev, and believed the governmental agreements ended the uncertainty regarding the Russian space station. Unfortunately, the fate of the space station rested now with Dr. Semenov. The Russian government agreed. If commercial funds could be found, the station would continue. In the year leading up to this last-ditch meeting, there had been foreign interest, some real, and some schemers seeking only publicity.

British businessman Peter Llewelyn received the most media attention. In April of 1999 this man who the newspapers described as having a 'James Bond' style, agreed to pay $100 million for a week long ride onboard the space station. The 51-year-old international would-be space explorer's business included waste re-cycling in Russia, an expertise I didn't know existed in Russia. Yet he was only the latest in a series of dubious commercial interest in the Mir. Energia had asked me to help in the discussions, but the whole situation smelled, excuse the pun, and I passed. Even from Washington one could tell that the size of the price tag for a week in space, the lack of information on his background, the eagerness of the "James Bond' businessman for publicity meant trouble.

By May this latest effort at commercialization was deflated. The Russian press began calling him "The Garbage King." His plans turned weird; he was raising funds he said for a private hospital. Then British reporters uncovered that he had been arrested for improper business practices in the United States. He began training but the promised funds never arrived and finally the head of Star City, where the cosmonauts train, booted him out. Scrappy Petr Klimuk, who runs Star City with an iron rule, had enough, and the ultimate insult from a Russian was voiced. He was, said one, "a profiteer with whom it's better to not have contact."

During the summer of 1999 there was a Canadian who gave some funds and then disappeared. And then there was one I did assist briefly, an American with ties to Wall Street who sought a more traditional means to keep the Mir flying. But that fizzled as well given the lack of time, leaving Koptev to proclaim that the Mir would be de-orbited by year's end. Semenov finally realized that there would be no serious help from the West because of the political situation. An intriguing offer from Semenov to use components of Mir to drastically reduce the cost of the International Space

Station had been rejected. So too a proposal to transfer all the scientific equipment aboard Mir to the new space station. It was all about politics, reasoned Semenov. "Technically - and it was recognized even by US specialists - (our proposals were) feasible. But as soon as the issue was raised to the level of US congressmen, it came to a full stop. Unfortunately, this confirms that manned space flight is still in the realm of big politics," he complained to a reporter.

"Why do you come now?" thundered Semenov, staring down the Americans.

"You come from a country that threatens my country," he went on, pounding his fist for emphasis. "You come when American planes, NATO planes, bomb a European nation not at war with you. Why do you come when no space companies from America will do business with us, when your Congress and President votes to expand NATO closer to our borders." The rage continued. "On my shoulders rests the issue of the safety to the world from the station de-orbit." Semenov turned and snapping his fingers ordered one of the waiting engineers, Anton Grigoriev, to produce a report. Then, through the young interpreter, Julia, he continued. I can't help but wonder what this young woman, standing with one foot in the new world and one here in the old, thinks of the effort to save the aging space station.

"You will see for yourself this report," thunders the General Director, and his gnarly hands trembled as he fumbled with the Power Point slides. "It shows there is a 4% chance of the Mir crashing onto your country. You can see for yourself the dangers. I stay awake at night, thinking how to save this station, how to de-orbit this station. I don't have the money for either option." He was telling the truth. Everyone in Moscow was explaining with regret this latest Russian paradox. The government lacked the funds to continue the space station; yet lacked the funds to bring it down safely. The Progress cargo ship belonged to Energia. Funds were needed, but how could one find the funds to throw something away?

For now, the only hope left seemed to lie with a Russian actor and his Hollywood backers. The world's largest, oldest and most complete space station's fate rested on whether the well-known Russian actor Vladimir Alexandrovich Steklov would be able to film movie scenes onboard the station. He was working with a Russian producer, Yuri Kara who had enlisted for financial support John Daly, the British producer of *Terminator* and *Platoon*. The film was called *The Last Journey*, and would involve a renegade cosmonaut who refuses to leave and insists he'll orbit the Earth for the rest of his days. Ground controllers decide to send up a woman to lure him back. Daly was filled with optimism when quoted by the Associated Press. "We will be hoping to get the right actors and actresses for the film," Daly said, expressing hopes of involving stars such as Robert De Niro, Gary Oldman, Sean Penn and Catherine Zeta-Jones.

One payment had been made; a second would come with the launch and the final after the conclusion of the flight. But there were problems in negotiations and the funds were as usual late in coming, and how could one continue a station based on a single commercial idea, and one where once again backers seemed more concerned with publicity than building a plan to keep the station alive.

This was beyond anything Semenov had ever wrestled with. The huge General Designer sputtered to a halt, looking over to his advisors who grimly nodded in agreement.

The World According to Anderson

Walt Anderson spoke. This successful investor had a few moments to show Semenov that he was serious, about the funds, about working with Energia and equally important, that he was not there as a representative of the aerospace industry, nor some scheming self-promoter. This was everything Walt had worked for-the reason he made his millions-to role the dice with the oldest and most knowledgeable space organization. "The reason," he began softly, "no one in America will come here to do business with you is very simple." And Anderson correctly stopped, waiting for the translator. Then he continued. "The aerospace companies have sucked you dry of your space technology, or as much as they think they can get, and now they don't need you anymore." The Russians registered surprise at this honest appraisal.

"We are here," began Anderson, "because we believe in the value of the Mir space station. We believe it is a beautiful piece of machinery, and I believe that you don't want to rely in the future on NASA for manned operations in space aboard the International Space Station." Continued Anderson, "everyone in this room understands what the situation is today. NASA and the aerospace contractors must remove all low-cost competition, and that is what they are doing. They spread lies about the Mir space station, they poke fun of the station and meanwhile they can't build the ISS without you.

"I believe that a large group of mostly European and third world aerospace companies would welcome the idea of undertaking commercial work on a space station priced reasonably, without the sort of interference from NASA we saw during the Shuttle-Mir era." Anderson then launched into a more philosophical discussion, the likes of which I'm not sure Semenov had ever experienced. He explained that all of his funds were kept offshore. How he didn't like government, any government. "I don't like your government; it suppresses people and doesn't let them unleash their own creativity. Your government pushes people down. You must be more willing to let your people here make their own decisions."

Semenov bore into Anderson coldly.

"And my government is just as bad, but in a different way," Anderson admitted in the same low tone. "They want to rule the whole world and impose their military might on you and every other culture."

This was not the normal discussion at this table. Or perhaps any table where sat Semenov. Semenov was thinking it all through. He looked over at me and seemed to shrug. Now it was his turn to toss the dice. "What do you want, Mr. Anderson?"

Walt's voice was even. Not a hint of excitement or worry. The pen twirling in his right hand stopped. "I want to buy the Mir, Dr. Semenov." He stopped for the impact. "Mr. Manber tells me that is impossible. If that is true, I want to work out an agreement by which we can take that Progress cargo ship and boost the Mir higher and then together create a company. This is a long-term solution. Not a one-shot deal. You can't keep the Mir in orbit with a series of one-shots. There aren't enough movie actors and Japanese reporters and politicians. Maybe you survive today but will die in March. We need to change the corporate structure of the Mir. Do you agree?

Semenov was silent.

"I've been working on a possible structure for a company. It's called "MirCorp" and the company will be based in some European country. Not in America and not in Russia. It will be made up of people from Europe and Russia and America and my plan is to have this company market the Mir on a commercial basis. That's what I want, Dr. Semenov. I want to work with you on a long-term basis."

Those words represented a large jump for Anderson. Tumlinson had wanted to simply purchase the Mir and not rely on Energia. That was an impossible situation. Energia and the Mir were one and the same.

Semenov turned first to his left and looked at Ryumin, who seemed a million miles away, thinking of something else. It was true that sometimes

Semenov made key decisions against the advice of his advisors and against even what he was thinking just an hour before an important meeting. Nothing was more important than what he sensed from looking into someone's eyes. It was the old way of doing business.

Semenov then looked at Legostaev, who nodded in agreement without turning his head towards Semenov. Then he looked across the table to me, and I too nodded my agreement. Again he shrugged at me, as if to say, "who the hell knows?"

Just two weeks earlier Semenov was still not paying attention to Anderson even after meeting Tumlinson. Finally, in exasperation, after speaking with Alexander Derechin, I had sent a handwritten fax to Dr. Semenov admitting that while perhaps Anderson might seem crazy to Energia, he was a pragmatic businessman and was ready to work with Energia as a partner. With just days to go before we were to leave for Moscow, Energia had asked for another report on Anderson. I candidly wrote that he was unlike any other U.S. businessman to have approached Energia, because he was coming to fulfill a dream and not as a mandate from the U.S. government. My report detailed how he dressed casually, that his views on politics were non-orthodox and finally, the report ended by saying that Energia had always voiced their desire to work with true commercial businesspeople. Walt might be the first.

Anderson then began talking about his real dream: building a module that would turn, like a Ferris wheel, providing gravity for the people living there. "This could be our hotel, for long-term visitors. They don't need to worry about the effects of gravity." I cringed. Walt had a number of long-term ideas that flew in the face of business realities. I was hoping they wouldn't be brought to the attention of Energia's senior management. Semenov politely expressed interest while admitting that probably it couldn't be done as part of the current Mir space station.

"But later, perhaps near the Station?" asked Anderson, oblivious to the technical difficulties. Semenov smiled and nodded.

Ryumin finally spoke up. In his deep bass voice he addressed Walt. "Why do you come now, why not a few months or a few years ago? Why now when we have no time to save the space station?"

"You would have blown me off a few months ago" honestly answered Anderson. "It's time to bottom-fish." The young translator Julie was at a loss to exactly explain what the foreigner had revealed.

Chapter 20: But No One Believes In Success

The two merchants of space had at long last met. The American capitalist and the Soviet General Director shared much in common and yet were separated by so many barriers. Each willingly fought long odds; each had collected a good share of enemies, each was short-tempered, yet each harbored aspirations that in other industries would be tantamount to romantic daydreaming. The pity was that they were brought together in a moment of crisis and the relationship could not survive beyond the space station that brought them together.

Anderson intuitively understood Semenov's concerns regarding the shortness of time and had prepared a rather extraordinary gesture. Before leaving for Moscow Walt had arranged with his European bank to standby to wire a huge payment to Energia while still in Moscow-if the discussions went well. After toying with the size of the possible payment, Walt settled

Energia designer Vladimir Syromiatnikov (far right) was always ready to explore novel space concepts. It was his department that undertook the tether work for Mir-Corp that was later blocked by the U.S. State Department. His specialists had earlier worked on the imaginative orbital light producing Znamya project. Here Rick Tumlinson, the author and Gus Gardellini are joined by two of his most prized assistants, Oleg Saprykin and Ali Botvinko. Photo: Gardellini/Tumlinson

on seven million dollars. Certainly enough to be taken seriously. Nothing had happened in the opening meeting to warrant sending the money-but then again, Anderson found Semenov more open than he had imagined and Semenov had seemed receptive to the Anderson view of the industry.

Over lunch Semenov succinctly explained what Energia could provide to this new company called MirCorp. There were three vehicles: two Progress cargo ships and one manned Soyuz which had been built under the watchful eye of Dan Goldin and the program managers at NASA. These ships were manufactured for the International Space Station. But the launch schedule continued to slip so the vehicles sat idle, aging quickly. Both countries caused the delays. The Russian-manufactured module known as the Service Module was late since Koptev and the Russian government had no funds for hardware.

On the American side there were major software problems from Boeing, which received far less media attention in the United States. Congressional hearings in 1998 and 1999 blasted the Russians while overlooking the American problems. Only a few experts, including the respected industry newsletter *Aerospace Daily*, commented on Boeing's technical challenges.

The delays had caused these three vehicles to remain on the ground. In another few months, the Soyuz would be too old to be safely used. In addition, given that Semenov had never received from Koptev the full funding promised for Energia's manufacture of the cargo vehicles, Semenov felt under no moral obligation to keep them within the ISS framework. NASA and Goldin later loudly protested MirCorp's use of these vehicles and many newspapers and commentators took up the beat that we had stolen something from NASA and hence the United States. But we did not. Russia eventually honored its space station commitments. Lost in the noise was also that NASA had never objected to using these same rockets for bringing the Mir down.

If we could work an agreement those three vehicles would be available for boosting the Mir space station to a higher orbit and employing the tether. That is what Semenov had to offer Anderson.

Semenov and his senior advisors left us after the meal to work the details with Derechin and his team. We moved into the smaller conference room, more appropriate for the give-and-take discussions that were needed. It was a bare room with little heat, as was true for many of the Energia rooms. No fancy presentation aids, just two long tables.

Derechin outlined the technical situation. The first Progress was being readied for launch. We had to quickly reach an agreement to allow time to change the cargo from all fuel to some fuel plus oxygen and food for a future crew. Then we would have to pay for a Soyuz, for a manned mission, and one more Progress to stabilize the station yet again. The twenty million dollars was not enough money to save the station, Derechin flatly stated. There would be costs for the workforce, for the preparations, for the oxygen and so on. Anderson froze when hearing this. Derechin paused, and then offered that if an agreement on profit sharing could be reached then Energia would contribute twenty million in services and hardware. That was a pleasant development in the long day.

By nighttime everyone was tired and cranky. It had been a long process. Rick wanted to inject bold and broad statements of philosophy, but Walt and I kept pulling him back, understanding with Derechin that a simple document was the most powerful approach. This was not the Declaration of Independence. The power would be in what we do, not in what we say. Walt had me lead the discussions along with Gus. It was quickly clear that Gus didn't understand all of the nuances, how could he? So John Jacobson and I ended up working the framework of the deal with Derechin and his chief deputy, Yuri Makushenko. It was in one sense an unprofessional situation: I was there out of respect for Energia as I was at that time no longer officially working for Energia, and I was negotiating for Anderson. In fact, it was the first time in several years that I had sat opposite Derechin. At one point Gus motioned for me to move to the center, directly opposite Derechin. Gus leaned over and whispered "I'm giving you the keys to the magic kingdom, bro."

The outline of a deal developed. First, provide money by the end of December to launch the first Progress in January to stabilize the station. Then work a plan to jointly develop the tether system and deliver that to the station, either via the Soyuz or the second Progress, by June of 2000. At some point there would be a second payment for the second vehicle. Then a third payment for the deployment of the tether experiment. There were huge issues hanging on the table, such as who would pay if the Mir had to be de-orbited. What price would Energia charge for operating the space station? What about liability insurance? How could we determine costs for a space station service?

Walt was insisting that MirCorp would be a partnership and partners never make a profit from one another. Could Energia even understand the concept of charging at 'cost' and explain that to future investors?

No.

There were so many issues and just no time. Normally, a contract of this size and certainly this historic magnitude would take a year to conclude. We had less than six weeks.

Sixty-Forty Partners

Walt was assuming he would take the majority share of MirCorp, since it was his funds. That was logical if the effort was in telecommunications or some Internet venture. But I disagreed for two reasons. First was political. The Duma would not stand for letting control of the space station slip from Russia, even though they refused to provide the funds to maintain the Mir. "I'll give them 100%," Anderson remarked sarcastically, his eyes popping wide open, "of the station underwater, if that's what they want."

The second argument was more within Anderson's business comfort zone. We were leasing an asset that had some worth. Pick a number. Five hundred million dollars. One billion dollars. How much would it cost to build, launch into space and maintain six school-bus size modules? Energia was contributing those assets to the project, plus part of the costs of the initial launch vehicles. Anderson slowly came around to understanding this was a political issue. This despite Tumlinson. Rick fought every argument of mine, not wanting to give the Russians any say and any control of the project. "Walt, what have they done in the past ten years but destroy the value of their own space station?" In some ways Tumlinson was right but that was not the point. The reality demanded foreigners accept Russian control or not do the deal.

Anderson grudgingly mused that he could yield as much as sixty percent if necessary. But honestly, I forgot to tell Derechin. Off in one corner I heard Yuri Makushenko suddenly propose they discuss stock ownership and why not a split of 50-50 as partners. Fortunately, he was speaking in Russian and so I had a few seconds. I rushed over to the table and cried, "stop" to the translator Julia.

Turning to Yuri I quickly explained in my poor Russian that Anderson had agreed to a 60/40 partnership. I asked Anderson and Jacobson for a "twenty-second time-out" as I called it, using basketball lingo. They agreed. Not professional at all but I didn't care. This was critical and once it was on the table as an even partnership it would stay that way. Then I went over to Derechin and explained. He was a little surprised that Anderson would agree, but immediately gave his approval.

This concern had been seconded by a Russian government official, one I had met at one of the receptions for Vice-President Gore and Prime Minister Chernomyrdin. The story he told that evening was difficult to forget. In the 1930's his Jewish father had sought to emigrate to America as a young man but was turned back at Ellis Island because of poor health. "So the difference between you and me," he summarized, "was the good health of your grandfather and the poor health of my father." Among a few trusted friends I would come to refer to him as Deep Throatskii, in recognition of the "Deep Throat" who advised the *Washington Post* Watergate reporters Woodward and Bernstein. As a holder of a position high within the Kremlin, he was able from time to time to provide some inside guidance, but I certainly pushed the limits during the pressures of MirCorp. Deep Throatskii had warned in a quick meeting in the hotel lobby late the night before that the Russian hardliners would never permit the space station to be controlled by foreigners. Sending a ticket holder into space is one thing- yielding control to the West for the Mir platform is quite another. Politicians in the Duma wanted some cover.

After a brief consultation with Derechin, Makushenko returned to the table and calmly proposed the new arrangement. He stated that given the proud heritage of the Mir, and its ownership by the Russian government, they were thinking of something admittedly unusual (he was hedging his bets in case I was wrong). How about a 60/40 split? Anderson hesitated. He mused aloud about trusting Energia and understanding the significance of the Mir space station. For that reason he would agree, subject to the understanding that the financial investors would direct the business direction of the venture.

"No problem," agreed a relieved Derechin. "Walt, you have our complete agreement on that."

By evening's end we were all exhausted. We had made extraordinary progress. Thanks to the unusual trust being shown by Anderson, we could state that the foreigners in MirCorp were secondary to the Russians in stock control. Ownership of the station remained with Russia, and the foreign company would lease the services of the space station. As they say in real estate, we would be responsible up to the "first coat of paint." On business issues Walt Anderson and the other investors would have the final say. On technical issues, the Russians.

Let's Calculate the Chance of Success On Paper Napkins

Walt was only staying for one more full day, and Rick wanted to show him some of the city. We went off for dinner without anyone from Energia, which shocked Anderson. Here he was prepared to sign a historic contract with Energia, and Derechin and his staff scattered as soon as we had finished our discussions. We were loaded back into the van and sent on our way back to the hotel.

Anderson asked me to invite Semenov to dinner. I refused. Moscow was now a place foreign to Semenov and the other officials of Energia. We needed to arrange something special, in a quiet hotel or private club and have a definite business function to the meeting. These were not men who would suddenly jump downtown and head to a noisy café. Ignoring my advice first he invited Semenov, who politely declined. Then Derechin, who also declined.

We walked from our hotel for about ten minutes along Prospect Tverskaya. It is the major artery leading up from Red Square, a huge thoroughfare filled with the best shops from Europe and even a few from the States and optimistically some Russian as well. I was slightly concerned about having Anderson in Moscow given that he had flown on his private plane, and had considered hiring some bodyguards. We decided that bodyguards would only draw attention. Everything was fine; Walt dressed like a tourist, not a prosperous businessman.

As we walked to dinner Gus and I fell behind the other three. Gus explained that men like Anderson use people as tools for their corporate gain. "With Anderson there are only two relationships, warned his friend. "You are either a tool or an enemy. There is little in-between."

We turned into Kamergerskii Passageway, a new pedestrian brick-filled street across from the old Telegraph office filled with good restaurants and cafes and even a Chinese Restaurant. Like everything modern, there was some good and bad here. The old street had been wonderful: a hodgepodge of antique shops, jewelery stores and second hand bookstores. Now there were benches and old-fashioned ornate street lamps. Very nice. But I missed the old jewelery stores where women in fur coats jostled one another to buy amber.

We went to Café Des Artist, a restaurant run by a Brit. After ordering drinks Walt came out with a question from left field. The entrepreneur asked us to guess our chances of success. Walt was dead serious. "You people here are my advisers, and Jeff, you know the Russians as well as anyone. I want to hear from each of you."

His method was equally unusual. Walt ordered us to write down on torn up pieces of napkin our prediction regarding the chance of success. The timing was strange. We were already in negotiations to keep a space station in orbit, against the wishes of the United States and many in the Russian government, and now he was asking our opinion be given on scraps of paper. It was vintage Anderson. Unorthodox in style and strategy.

I asked Walt to define success. "Success meaning we can take control of the space station, market it as we wish, professionally and honestly, and attract customers and investors to maintain the station." It was of course the question of the day, and I had given it a lot of thought. Honestly, I thought there was little chance of success. Not because of the Russian side. We could work with Energia and solve the political issues from a Russian perspective. But where the hell would we get the money needed to operate the station? Who in their right mind would invest in the Mir space station, invest in Russia, and invest against the United States?

How much does it really cost to run the space station? If the tether project worked, we might have some chance of structuring a financial package. But if the tether failed, we had months, not years to find the money to pay for more rockets. I had decided it was best to keep silent, since that wasn't my role here, not for Energia or for Anderson. I had been asked by both parties to help bring about an agreement, not for my opinion on the merits of the deal. Walt was now changing the equation.

Briefly there was the thought of lying, but Anderson is a straight shooter and deserves nothing less from those around him.

Walt pulled John Jacobson's prediction first. I think John had written there was a 50% chance of success. Fifty-fifty. That surprised me. I thought I was the only pessimist. Gus was next. His said half of one percent. Under one percent! No chance at all. "And I'm being optimistic," he blurted out.

Mine said 30%. Seventy per cent chance of failure.

Rick also had 50%. Could go either way.

So everyone surrounding Walt believed that the effort to market the Mir could not achieve commercial success. Everyone sitting there in Moscow believed the odds were against us. Either the Russians would swallow us up (Rick and Gus thought) or there would be no customers or investors. (John and myself).

Gus tempered his negative opinion with a passionate statement that "never in my life have I so wanted to be wrong." His hands were flying and his whole body language was filled with energy. "Prove me wrong Walt. Prove me wrong Rick. Run this space station and show everyone the arrogance of NASA. But you can't. You'll be crushed trying." Then Gus got even me emotional. "The government will kill you," he shouted, his finger pointing right at Walt. "You will be a dead man if you do this project."

Walt looked past Gus. My feeling was this was not the first time he had heard this passionate warning from his friend. Anderson then asked my opinion of the Russian side of the project. I stressed the need to create a package that appeals to the most nationalistic side of Russia-that was our only hope of attracting the political support within Russia to stand up to the State Department and Dan Goldin's NASA. The more business oriented Russians had surrendered years before to the idea of American supremacy, so they would not support poking NASA in the eye. That is a harsh statement but true. The reformers wanted a role in the NASA space station amongst a range of international opportunities. But I concluded by saying all signs suggested Energia was ready to honor the agreement, once Semenov signed.

Walt smiled his tight smile. "Well," he said, "I stupidly asked for your opinions and you all gave it to me, uh, I guess I have to say thanks?"

Gus continued his tirade. "They will tear you up, Walt. Either the Russians will screw you out of the money or the Americans will go after you on tax fraud or some shit like that. Who the hell are you Walt? Who the hell are any of us to sit here and think we can take control of the station."

Rick also feared working with the Russians. His view was unchanged. "Let's buy 'em out and use them like mechanics on a plane. For money we can control the whole situation." This wasn't John's industry so he had little to say.

To me, the Mir was more organic. You could no more take the Mir 'out' of Russia than remove the London Bridge and use it as anything other than a tourist attraction, or take the White House from Washington and consider it a center of power. There were thousands of people throughout the Russian space industry working every day, even when it was unmanned, to keep the Mir in orbit. The unseen ground army was from the parade of organizations I had witnessed during the Energia anniversary event in 1996. Their loyalty was based on more than money, it was a pride that had better be part of the equation or Walt will sink as fast as others that have attempted to take control of some Russian asset-in other words, before even starting. I didn't say this, but success would be if MirCorp was simply allowed to take control of commercial operations.

The talk moved on to Walt's dream of putting up a module that would slowly turn, it seemed his vision was straight from the old science-fiction pulp novels of decades ago. Rick thought Semenov had approved the idea. I let the conversation drift away. My thoughts were instead on Walt. He was prepared to spend millions of his own dollars and each and every person around the table at Café Des Artist didn't think the odds were in his favor.

The owner joined us. Mike was an older man, with dark hair and a delightful British accent. He was dressed as one would expect, in a blue blazer and gray slacks. We had met him on the earlier trip and given our unusual mission he certainly remembered us. With excitement Rick explained how the discussions were proceeding. Mike was polite and generous, buying us a round of cognac. He was also cautious. This was a land of skepticism and we were round-eyed dreamers. Walt asked if there were local sources of capital that might want to work with the station, given that it was a source of national pride, yet the company would be located somewhere in Europe. Mike's answer was simple: "there are Russians far wealthier than you Walt-you just have to get to them." He added correctly that for most of them, Russia was the last place they wanted to invest.

"Isn't there any sort of patriotism?" Walt innocently asked. Mike shot him a look of surprise and simply answered "patriotism in Moscow begins and ends with your own bank account."

The extraordinary day ended with the obligatory visit to Night Flight, the club where foreign businessmen mingle with beautiful young Russian girls. The drop-in provided the final data point on Walt's perspective on both himself, and the ethics of capitalism. Within moments of arriving Rick and Walt were in a heated argument over prostitution, with Walt believing sex for money was unethical.

Rick stared at his 40-something multi-millionaire benefactor, dressed in his drab tan overcoat, worn baggy jeans and old shoes, who had, waiting for him in another European city, two tall and beautiful young women. "Why the hell are they waiting for you," bluntly pushed Rick. "For your good looks? No, because of your money and your private jet and your expensive wine." Anderson genuinely fought back. Did he really think that his girlfriends were not there in part because of his lifestyle?

The argument between the two lasted all the way back to the hotel.

Anderson was naïve about social and political issues, but could focus with a Zen-like enlightenment on how to make money. This project would take both-his ruthless ability to find profit and yet articulate a new vision for manned space operations. I'd be satisfied if we met only one of those objectives.

Chapter 21: Anderson's $7,000,000 Phone Call

The next two days would be a race against the clock.

Walt was insistent on leaving on his planned schedule even with major issues on the table. One sticking point that emerged was how MirCorp would handle the Russian space research organizations that had long relied on their space stations for research. Derechin had raised the fear that there would be deep opposition in the industry and hence in the Kremlin if Mir-Corp behaved solely as a commercial platform. "How would it look," suggested Derechin, "if the Europeans or Brazilians are using the Mir space station, but the Russian scientists cannot afford this?"

Anderson was unimpressed. "Investors are entitled to a return on their investment. We can't give things away." And then, in a jab at Energia's behavior over the past decade, the would-be investor added, "just like you did for NASA."

Anderson's dramatic gesture to wire seven million dollars to cement the MirCorp negotiations flopped. The Russians just didn't believe the funds had been sent during lunch via his cell phone and the congratulatory toasts were forced. When the funds did arrive Energia panicked since without a written contract the money could be confiscated by the authorities. Photo: Gus Gardellini

Another problem was Walt's insistence on using Bermuda as the corporate location for MirCorp. Energia was unfamiliar with this offshore jurisdiction and feared having to quickly learn the intricacies. Anderson was surprisingly unresponsive. "I'm the businessman; I will take care of the financial structure. You take care of the rockets."

At one o'clock we broke for lunch. Semenov, Legostaev even Ryumin joined us. So too Zelenschikov and several other senior Energia officials. It took place in one of the nicer inner rooms, just off of the historic conference room. There were the obligatory toasts and then the lunch broke down into smaller groups, with the Russians chatting among themselves. Everyone seemed relaxed and satisfied at the extent of the discussions, though we would have to have a formal review once Anderson left Moscow. While finishing up the breaded fish Anderson turned to me. "Are we close enough on the issues?"

"Close enough to what?" Was my perceptive answer.

"Close enough to just go ahead and send the money. That would make them comfortable, right."

This was exciting yet also worrisome. Lunch was not a good time to talk business with Semenov. Serious business, like sending the funds, should be done quietly, in a side room, with few others present. But I dutifully answered. We were still far apart on questions of production costs, payment plans and some other issues. We had agreement on the basic principal. MirCorp would control the marketing of the space station starting sometime during the manned mission in the spring of 2000, the exact time in dispute.

Anderson nodded and then addressed Semenov. "Dr. Semenov, Jeffrey tells me we are close on the major issues. I know you are used to dealing with NASA and big aerospace companies who never want to pay you for your services. I want to pay you as a partner pays another partner. That means quickly." Semenov had no idea where Anderson was heading with this little speech and was hardly paying attention. The main translator, Julie, was eating her well-earned lunch and this was not the time for working. As usual, Anderson stubbornly continued. I asked Legostaev to translate.

"Translate what?"

I explained that Anderson wished to make a payment to Energia right now, before the lunch was over. "That's nice," replied Legostaev and he continued eating while muttering something to Semenov. The rest of the Russians were talking and having their coffee.

Walt pushed on. "I'm going to make a phone call now and wire the first payment to Energia." Between bites on her own meal Julie dutifully translated.

With that Walt picked up his cell phone and reached his banker somewhere in Europe. "Yes, this is Walt Anderson, I'm ready for the wire. I DON'T CARE IF HE IS IN A MEETING" His raised voice caused some to look over. Julie had resumed eating as she assumed Anderson was making a private phone call. Gus was grinning. John was grinning. Rick was grinning. And I sat there stunned by the wasted opportunity. Finally the banker was on the line. Anderson continued his dramatic moment. "Yes, this is Walt Anderson. The password is "Mir is for peace." Yes. Good. Thank you." And he hung up. "Well," he informed Semenov, "you will receive the funds in several days. It takes at least three business days within Europe" and off he went for several long sentences explaining to Semenov how bank transfers worked.

Legostaev said in Russian, "Mr. Anderson says he has sent us some money."

Seven million dollar had been sent to Energia. Without a contract. Without a signed agreement. Over the break for lunch.

This should be the end of the story. Or mark the beginning of trust between the Russians and a high-flying Western entrepreneur. Indeed, Neil Bourdette, writing in a front page article on MirCorp in the *Wall St. Journal* on June 16th, 2000, highlighted this phone call as a dramatic moment.

"The cash strapped Russians were days away from letting Mir burn up in a fall to earth, wrote Bourdette. "One of the Americans, Walt Anderson, a camera-shy investor who made a fortune in 1992 selling a U.S. regional long-distance company he had started, says he sensed the need for something dramatic. He pulled out his cell phone, punched in a long series of number and told his bank to wire $7 million (7.4 euros) to RSC Energia. 'You should have the money in a few days,' he told Yuri P. Semenov, Energia's president, according to three of the people in the room."

The call changed the course of the Mir, and maybe the history of the commercialization of space.

It sure sounds like an incredibly dramatic moment. It was not. It fell flat since not a single Russian believed or understood what had happened. I jumped up and made a toast thanking Anderson. Ryumin just laughed. I looked to Victor Pavlovich for support. "It's nice," he again agreed and resumed digging into a chocolate éclair.

Life is so much more mundane than we would like.

After the Seven Million Dollar Phone Call

We continued after lunch to negotiate the details. Anderson was a little perplexed at the lack of response from Energia. When the conversation turned to determining production line costs of the Soyuz and Progress, he voiced his irritation at the slowness of everything around us. He interrupted a technical Energia presentation saying how "if you guys read all the signs on the road when driving from point A to point B, you will never get there. I will tell you how to get to point A and B without reading the signs".

Everything we were hearing in the afternoon confirmed our belief that Energia did not really think in terms of real costs. It was more either a barter perspective for raw products, or a mentality of how-much-will-the-customer-pay kind of thinking. The Russian government had even less of an idea and viewed commercial organizations such as Energia or Khrunichev as their private fiefdom's to siphon off needed funds or perform needed services at less-than-market costs.

Between being squeezed by their own government and NASA's philosophy of condemning commercial companies, and the U.S. State Department's insistence on keeping all things political, it was no wonder that the Russian industry was suffocating even though it produced the finest space hardware at the lowest cost.

We kept insisting to be told the actual costs of the manufacture of a Soyuz and they kept refusing. We told them they could even fudge the numbers. After all, if a NASA space shuttle flight cost $500 million, we don't need to know a Soyuz cost $8 million. Tell us $20 million. It is still a major step towards genuine space commercialization. You can sell a seat to space to the private sector for $20 million, but not a ride for a $100 million.

Every few hours an Energia person would approach me with the news that they had received no funds and no confirmation that the funds had been sent. They were begging us for a wire receipt-but Walt kept saying this is part of the trust. I think he didn't want the Russians to know the details of his account and the other information found on a wire receipt. So much for trust. The next day there were still no funds. Now the Russians were getting nervous. Word had filtered out through the top level that the American said he had wired millions to them. Most of the Russians were gloomy, as if they felt a joke was taking place at their expense. "Pay for that old station, they thought we sent the money," and we would all laugh in some London pub and drink more beer.

We reached one breakthrough in the discussions, compromising for the first year by paying a flat price for an array of goods and services. In that manner they didn't have to tell us the exact costs and we could show the competitive advantage over NASA to new investors.

That night about seven o'clock my hotel phone rang. It was Alexei from Derechin's staff. "Jeff, we have received the funds!"

"That's wonderful news." I was relieved.

"No, Jeffrey, it is not." Shouted Alexei over the poor connection. "You see, we have no signed contract and therefore it may be illegal for us to have these funds. What should we do?"

That is the end of the story of the seven million-dollar phone call. We left Moscow with major questions regarding Bermuda regulations, prices for launches, structure of the partnership and a timetable for future payments still uncertain. Worse, Energia was in a panic over the unauthorized money. We found the whole situation bemusing. There was no party. No celebration. We were mired the last day in technical contractual questions which gnawed at Anderson's trust in the entire situation. Anderson thought he had a partner now. He did, but in a Russian sense, not Western. Energia would do as agreed and that was far more than normal in Russia.

There was one more culturally embarrassing moment. Anderson visited with Legostaev and Derechin to discuss his vision for working the payment milestones over the next few months. Walt made clear that he had worked hard to send the money now so that the Energia workers would have some funds before the holiday season. "It will make a lot of people happy here, isn't that right?" Walt asked Legostaev, pleading for a hearty thank you. Legostaev demurred on answering.

Later, I turned to Derechin. "Should I tell Anderson that Russian Christmas is after the New Year?" I chided. He didn't laugh. Even worse, despite my briefings, Walt couldn't grasp that the bulk of his funds-no all of the funds- would go straight into the production line for materials and workers, and not to the staff working with Derechin. As we climbed into the van to head to the airport all that was certain was that as Semenov had promised the Progress cargo ship poised on the launch pad in Kazakhstan would now be stocked with food and oxygen and supplies in preparation of re-opening the station. It would be launched to boost the Mir, not destroy it.

We had changed not only the current situation in international space cooperation, but implicit understandings between the Russian Federation and the United States.

I turned to Anderson and empathized that never had a moment for celebration been so squandered. He just snorted. "I know the Russians are celebrating without us. They don't want to act like our partners just yet, but that $7 million will make a lot of people happy." Jacobson jumped into the conversation. "Walt, you just sent seven million dollars. This is mind boggling" His voice rose to a high pitch. "It's unbelievable."

Rick agreed. "It's the one thing Russians understand. And that's money. So they understood what you did. They've got to appreciate that."

"Yeah," Walt said softly, a grin on his face. "I guess we did it."

Anderson looked over at me as we bounced along in the van. "Look how serious Jeffrey is! C'mon Jeff," he teased, "It's not that bad. Look at the situation. We have at least several months to show the world that the Mir is a good space station and that NASA is inefficient and working with the Russians is a good thing. And everyone in the American aerospace industry wishes we all would go to hell."

Then Walt actually laughed.

Chapter 22: NASA's Worst Nightmare

At the end of November Mir's perigee was now down to 331 kilometers, still falling faster than anticipated.

Seven million dollars had sure gotten the attention of Energia's management. One individual, one wealthy individual hell bent on ramming through a new way for space exploration, was capable of distorting policy between the Russian Federation and the United States. Semenov had long wanted to meet an American like Walt Anderson-willing to behave like the American businessmen celebrated in our media. Willing to take risks, willing to build a market from the ground up and not so incidentally, one respectful for the path Energia had taken under Semenov. The same money used to buy a ticket in space would not have had the same ramifications.

The signing between Walt Anderson and Yuri Semenov. (Valery Ryumin is in the background). The MirCorp agreement remains one of the more interesting in manned space and in the recent history of Russia. It allowed a foreign commercial company to market fully a national Russian icon. The arrangement was supported more by the Russian nationalists than the Western reformers who worried about displeasing NASA. Photo: Gardellini/Tumlinson

The problems resulting from the Anderson plan were difficult to even contemplate. Space Station cooperation was more than a space issue. It involved State Department, the Pentagon, issues of missile and nuclear non-proliferation. Treaties were in place not only with Russia but also with Japan, Canada and the European Union. The cooperation between the Russian Space Agency and NASA had been implemented by Vice President Al Gore and Russian Prime Minister Victor Chernomyrdin. All assumed Mir was coming down. This was technically not accurate and resulted from NASA's refusal to look beyond Koptev for understanding the situation.

Any intrusion on the American perceived reality would be a personal black mark for Dan Goldin. On the flight back to Washington I was thinking how Goldin needed to drop into Energia with a bottle of vodka and two shot glasses, sit down with Semenov and come to understand his concerns and seek to embrace his vision. But he wouldn't and didn't, and now NASA's most cherished project in a generation was about to be threatened by two men, one the consummate insider pushed to the outside, and the other the ultimate of outsiders.

Now I had a personal decision to make. Whether to disassociate myself from the impending battle or become fully involved. After the meetings were over Anderson pulled me aside, and with this big build-up, thanked me for my help. He then asked me to take over as head of MirCorp.

I immediately declined. Shocked he turned to Tumlinson. "I thought we had taken care of this, Rick." Tumlinson began sputtering. "Ah, he's just playing hard to get. He wants this job real badly." My hesitation was no ploy. For the past year I had been working with a company named Astrovision that sought to provide the world's first moving real-time color images of weather patterns, clouds, hurricanes, and even night traffic.

NASA's Stennis Space Center in Bay St. Louis, Mississippi had asked me to assemble a management team supporting the founding engineer, Dr. Malcolm Lecompte. In just ten months we had brought in a first-rate management team, the first major customer and had just landed the first round of financing. I was enjoying myself. No routine late night phone calls, no trekking to Europe and Moscow and back in a week, no mess of cultural confusions, nor being socially ostracized by Washington insiders. Nor was I alone in that department. Around this time I had lunch with Charlie Chafer, who had worked with Deke Slayton to launch the Conestoga I in the early 1980's. Conestoga I was America's first commercial launch and it caused Bob Brumley and others to create a space division within the FAA to streamline the process of launching without NASA. Charlie was now working a project sending cremated ashes into space. The business was doing well, especially in Asia. He too was suffering from the disdain of the policy crowd. "I'm one of the few guys that can pay for these smaller launches. I have customers, I have demand. But no one in the space industry takes it seriously," he admitted.

I consoled Charlie by explaining what he was doing was called commercial business. But I understood how he felt.

Anderson called me several times in the following weeks asking me to reconsider. Tumlinson called and whispered that I was "the man" because I had taken him to a club in Moscow where you were frisked for guns upon entering. "That's just the kind of guy we need for this position," Tumlinson said in all seriousness. "Someone who knows space exploration, but can kick some butt." For several weeks in November I was torn. The opportunity to market an existing space station was unprecedented. Yet, freed from working with the Russians was somehow liberating. Making matters more difficult was an episode that had just taken place. During a trade show sponsored by NASA down at the Stennis Center, Dan Goldin had been overjoyed to discover me working for a NASA supported company and even raised my "escape from the Russians" with Senator Trent Lott of Mississippi, the then-Majority Leader in the Senate. "Not only is this new company Astrovision commercial," enthused Goldin, "but we have rescued the Chief Operating Officer from the clutches of one of the worst organizations in Russia." Lott had not the faintest idea of what the Administrator was saying. But I understood. Here the paradox of Dan Goldin was on exhibit once more. He was genuinely pleased I was supporting the commercialization of a U.S. government function: weather satellites. Yet this same Administrator spent so much of his time beating back another private company from its path of commercializing manned operations.

Finally, Anderson offered the most powerful incentive. If I didn't take the MirCorp offer, I would kick myself watching someone else attempt to create a commercial manned exploration company. That was a strong argument.

Alexander Derechin was in Washington on NASA negotiations. We met early one Saturday morning, before he headed home. Anderson, I learned, had cleared his offer with Derechin, and Alexander told me Semenov had also given his approval. Derechin was honest in his thoughts. He too was divided on the opportunity. Alexander was concerned about the personal risks. True to fashion, we discussed my situation not in his hotel room, but outside on the lawn where there was far less chance of unwanted eavesdropping. "You will never be able to work for Boeing or NASA or any other major U.S. company," was Derechin's first warning. That made me laugh.

"Alexander, I never have worked for them and now is not the time to begin."

"You will make many enemies." He sighed. "So many companies want the Mir down. You understand. We are a threat to so many people."

I did understand. Why fix up the Mir for hundreds of millions when aerospace contractors could earn billions on a new station. It was simple arithmetic.

"And chances are you will fail, and that may hurt your future career," warned Derechin. With false bravado I explained that in America we are not afraid of failure, it is part of the entrepreneurial risk.

"And NASA will see this as, what do you call it, their worst nightmare."

What about Semenov, I wondered. Would he let foreigners take control of his beloved space station? Yes or no? That was a tough question for this pretty November morning in Washington. The sun was warm as we sat on a bench in Old Town, just outside of Washington. I asked Derechin whether Energia would watch my back in Russia.

"Yeah, sure," replied Derechin quickly. "But remember we have no money."

Not having Walt's personal phone number I left a message on his voice mail, suggesting we meet and see if we could come to terms. There was one more point gnawing at me.

We met at Rumors, a large and busy café in downtown Washington catering to quick bar food. I was learning this was typical Anderson. Sure, he liked three hundred-dollar bottles of wine, but he also liked his fast food. Walt ordered a cheeseburger and I had a salad, the better to eat while having a serious business conversation. That afternoon I witnessed one of the few ostentatious displays of Walt's wealth, albeit in a typically strange manner. Anderson had parked his car right in front of Rumors, clearly in a no-parking zone. Apparently, he did this several times a week. And every time his car was slapped with a hefty parking ticket by the same ticket lady.

I outlined for Walt some of my preconditions and concerns. I needed to be honest. We had to have a strong business base in Russia. This could not be seen as taking the best of Russian technology and exporting it. We talked about several personal issues, and finally I raised the very reason for the lunch. "This is kind of a sensitive point," I began. Walt looked up from picking at his fries. "Let's just say we can't raise further funds, or we don't bring in any customers. But we show the world that it was folly to bring down the Mir space station and that it is possible to commercially run a space station at a cost far lower then NASA 's boondoggle. Is that a success for you?

Even if you are out millions of dollars?"

Anderson looked down at the messy plate. I heard him say "yeah." He said it so low. I understood it hurt this entrepreneur to admit a motivation other than monetary. But I needed to know. I was more confident that we could get across some serious questions about the conduct of NASA and space exploration as well as relations with Russia. Provided with adequate money we could front-load everything and create a buzz about the space station from the moment we signed the lease, and worry about the tether and customers later.

Then I heard myself agreeing.

Events were moving faster and faster. I telephoned Derechin and told him of my decision. There was a pause and then the expected "congratulations to both of us." He always said that when we had news that meant long hours of work. Then he said the equally expected, "Fine. Let's get to work. We have so little time."

We set a date of mid-December for the first board meeting of MirCorp. It was Anderson's task to establish first a Bermuda-based company and then a daughter company in a European location. He promised this would all be taken care of in time for the board meeting. In Moscow Derechin had begged him not to locate in Bermuda. It was difficult to receive authorization for an offshore company from the Russian government. "We have so little time," Derechin lectured the international businessman. "Keep it simple." Walt ignored the warning.

I suggested to Anderson a board of five officials. Three from Energia or the Russian side and two representing Anderson. This was part of the firm conviction that the project was impossible without the strong and committed support of the most conservative elements of the Russian society. Energia had quietly asked for this, and I had tested it with friends up and down the Russian government in several late night phone calls. Still, it was like punching into a fog, since no Russian wanted to discuss the issue fully on the phone. I well understood that the Mir would be in the ocean before foreigners would take control. This was a delicate game of perception-we had to operate as if the station was ours while at the same time letting Energia report to the Duma and the Kremlin that they-and Russia-had not lost control of the world's only orbiting facility.

Walt was skeptical at first. "Let me understand. I put in $20 million and they get the majority share of the board? And they get a majority share of the stock?" It was a tough pill to swallow. He was saving the space station and here he was being told to take the minority position. Was this my idea or Semenov's? He didn't ask. I knew with every such suggestion Rick was growing more convinced that I was looking out only for Energia and the Russians. It was a sticky situation. This was going to be a difficult couple of months.

By this time I was looking beyond the Yeltsin Kremlin. There were persistent rumors that a far more Russian nationalistic program would take control from the weakening Yeltsin. This was all uncomfortable for Anderson. Here was a guy who refused to meet with any government official because of his distaste for the military and he was being told to structure his investment to satisfy nationalistic Russian sensibilities. Anderson thought the issue through out loud. This was a habit of his. He realized that if Energia wanted to "screw him" they could do so with or without a majority share. All Energia had to do was declare the Mir unsafe and the project was over. That was the primary demand from Energia. All safety issues, all questions relating to the operation of the space station, rested with Energia. If they said it was over, based on technical considerations, the project was over. "Tell Derechin I agree as a sign of trust. In return, I want a strong direct commitment from them that the Western partners make the non-space, non-technical business decisions.

"I won't tell them how to build a rocket and they won't tell me how to raise capital," was his oft-repeated mantra.

Once Derechin reconfirmed their willingness to let Anderson run the business of the station, and Energia the day-to-day operations, we moved ahead with the structure of MirCorp. Semenov proposed himself, Ryumin and Derechin as board members. Ryumin was important since he was respected by the cosmonauts, as well as being politically astute. Derechin's inclusion was testimonial that since I had 'brought' Anderson into Energia through Derechin, this was his project within the space company.

Anderson proposed himself and a new investor. This was wonderful news. His name was Dr. Chirinjeev Kathuria and he had apparently agreed to put several million dollars into MirCorp. That was all I knew. I finally spoke to Kathuria early in December and we would meet in Moscow. He came from the Internet community and was enthused about the audacity of the project. "This will be the most important project in my portfolio," he gushed. I explained to Derechin that it seemed that Anderson could reach out to his friends and colleagues and raise tens of millions of dollars through his own network.

Derechin was skeptical there were more people in America like Anderson.

Chapter 23: Derechin Is Worried

On December 15th Mir's low point in its orbit was recorded at 320 kilometers. At this rate, assuming the continuation of the current solar flare activity of the sun, the space station would be too unstable for a docking by spring. The space station would not be fully destroyed during re-entry. Its five modules together weighed more than 130 tons and if the reentry was not controlled exactly, the results could be catastrophic.

December, 1999

Back in Moscow for the final sprint in negotiations before the critical board meeting that would formally tie Russia and Anderson together. John Jacobson would join me in a few day's followed later by Anderson and the new investor Dr. Kathuria.

Alexander Derechin and Walt Anderson discussing the production line schedule of Soyuz and Progress rockets for both International Space Station and the Mir space station. The Russian company was by this time weary of supporting the promises of the Kremlin to supply flights to NASA at no cost and yet having to fight NASA over the right to earn profit for space operations. Photo: Chirinjeev Kathuria

We were staying at the historic Hotel Savoy and it would become our de facto office for the next few months. The hotel became my second home, and management worked to make us all comfortable, though they refused one critical request. I discovered from a confused taxi driver that the hotel name was new. The Savoy had been known for decades as the Hotel Berlin. This was the hotel that Lee Harvey Oswald had stayed in when he arrived in Moscow and subsequently refused to leave. The eventual assassin of President Kennedy holed up in one of the rooms until permission was granted for his demand to live in the Soviet Union. Despite my incessant asking, and eventually begging, the manager never disclosed which room had been Oswald's.

Before arriving in Moscow I scanned the Internet to learn if news had leaked. I was pleased that Rick had resisted temptation and said nothing to his wide circle of reporter friends. Rick had also begun briefings with U.S. government officials regarding export of the tether to the Mir, and they too had said nothing to the media. In my agreement with Anderson, Rick would not be involved in the operation of MirCorp, except as a very welcome adviser. Instead, he would lead the effort with the tether. The tether was the key to any realistic hope for keeping the station alive beyond a few months or a year.

Trouble began brewing as soon as I arrived. The Associated Press and Interfax, the Russian news agency, were reporting that the Duma had authorized funds to preserve the Mir. Tumlinson was on the phone immediately.

"They're screwing us! They're lying, man. What the hell are they doing?" It took me some time to explain how the Duma always authorized funds and nothing ever happened. But at the same time, the confusing news would be harmful to our efforts to find investors. There were also news reports that the Russian actor Steklov continued to insist there would be money coming soon to the Russian Space Agency and Energia to shoot that film on the space station. I would have to ask Derechin the true situation.

My return to Moscow and the implications of the call from Tumlinson made me feel pretty low. It was the first time that I had turned around from Washington and returned to Moscow after a brief absence; not enough of a separation to shake off the heaviness felt from spending too much time in the Moscow winter.

Meeting with Derechin did nothing to make me feel better. We were swimming in problems. Events were moving too fast for the giant space company. Overseen by one man, Energia was a deliberative body of competing fiefdoms. There was no time here for the usual discussions and jockeying among his advisors. Even more ominously, Derechin himself was uncomfortable. He was nervous about the Bermuda registration for MirCorp. He was nervous that the American side was writing the board resolutions with an attitude that seemed too casual to him. Putting it all together, Derechin was concerned that Anderson was hiding something.

Anderson's funds were coming from his offshore venture capital firm "Gold & Appel." Energia asked for the corporate records, an annual report and other background information. This was refused. Needing Walt to understand how this could derail the project, I called him from Moscow and he returned the call about midnight, waking me up.

Half asleep I gave him a perspective on the problem. Energia for better or worse, obeyed all Russian regulations, and certainly government officials would want to know as much as American government officials or Bermuda authorities regarding Gold & Appel. Energia's obedience to governmental regulations was also the bad news. It meant we had to proceed in a slow, careful manner. Semenov's advisors, from the national security side, from the operational side, from the financial side, were all asking for the proof that Anderson wasn't involved in some illegal or unsavory activity. "It's called trust," shouted Anderson, before hanging up.

The mystery surrounding the source of his funds was haunting us on many fronts. Not only had it slowed the approval from the Russian government, but Bermuda authorities were not amused. Bermuda does not approve a company that has stockholders with "blank shares." Throughout December there was a running battle between Bermuda regulators, our Bermuda lawyers and Anderson. They would not enter Gold & Appel as a shareholder in MirCorp. Anderson yelled and screamed and fumed. The answer was still no.

John Jacobson had joined me in Moscow for the negotiations. In the time since our first meeting together I had grown to like Jacobson. A tall man with a slight stutter, Jacobson had proven an important anchor to the team. Gus wanted no part of the project. Rick had an aversion to spending much time overseas and was not suited to the nuances of negotiations. Walt was off running his other companies. So John and I had to move the project forward. Jacobson witnessed firsthand the schizophrenic nature of Energia's management. They were desperate for funds to save the space station, but nonetheless slowed down the MirCorp project out of the caution that was embedded in their organizational DNA.

The latest problem waited for us in Derechin's office, where the current round of negotiations was taking place. Just a few weeks earlier Anderson and his lawyer, Jim Dunstan, had discovered one of the most amazing characteristics of Energia. The company that launched Gagarin and Sputnik, the company building the first two modules of the International Space Station, the company participating in the multi-billion dollar Sea Launch project, the company that was commercially launching European and American astronauts to space, had one lawyer for international business agreements.

Yes, one lawyer.

He was an engineer who worked all week and nights and weekends studying the contracts proposed for international projects. He had not taken a vacation in many years and frankly, may or may not have time for studying the MirCorp documents in the next several weeks.

His name was Viacheslav Vassiliev. I first met him during the early days of the Sea Launch venture. Viacheslav was a short man with an open, very agreeable face and wide eyes. One morning, in Seattle, I walked outside of our hotel and the only other person on the street was Vassiliev. He was walking into the hotel with flowers. Vassiliev explained it was for a Russian-American woman on the Boeing team. "You see," he explained in that strange singsong accent of his. "It 'tis for this woman and not for you and so I cannot stop and speak to you."

Vassiliev was worried. He and hence Energia had no experience working within Bermuda. For the Sea Launch project he alone had mastered the regulations of the Cayman Islands. This would be more complex and we had less time. Anderson just stared at me when I had explained the situation. "They are a rocket company," he began. "They do not know about Bermuda business regulations because they are a ROCKET COMPANY. I don't know about manufacturing rockets. I will not learn rockets for this project. That is their job. Tell them to trust me. I will establish the company." I told Anderson it wouldn't work. They would want to understand everything before moving forward. "Have them hire a good law firm in Moscow to do that sort of drudge work," he next suggested, calming down.

"Can't. Energia doesn't hire lawyers."

This was too much and I understood. Now he was speaking, not yelling, but speaking forcefully, as if to a foreigner not understanding what he was saying. "Have-them-hire-lawyers-with-the-damn-money" he paused for a deep breath, "they-will-get-from-me."

"Can't. No can do. Not possible. Won't happen."

Dunstan scrambled to save the situation. "Let's set up a meeting between this Energia guy and Bermuda lawyers." That too was a problem.

Vassiliev would need a visa to travel and that takes a long time. I think this was the exact moment that Anderson began to understand the complexity of working with Energia and with Russia. It's never a pretty sight when it really hits an American how different a Russian organization was run from every other organization the successful businessman had ever worked with, from the States to Asia. No, this was definitely the ultimate cultural clash, bringing together the freewheeling Anderson and the careful space engineers of Energia.

Vassiliev had studied the web site of the law firm Anderson was using in Bermuda. The site outlined key provisions for establishing a Bermuda company. It seems the first step is the formation of a local board, with each member being given one share of stock. This Bermuda board meets once and votes in the board nominated by the founders of the company. So in this case they would meet and vote in Semenov, Derechin, Ryumin, Anderson and Kathuria. "How do we know this is not a trick?" asked Derechin. "How do we know these men will do as we ask? What happens to them after the board meeting?"

It was a routine move. Not worth giving any thought to and at the very most, ask Anderson, get his answer and move on. This was not acceptable to Energia. Derechin wanted some sort of guarantee that there would be no tricks from the Bermuda founding board. Anderson was now angry. "I thought the idea of forming MirCorp was to teach these people how to structure a company to raise funds," he fumed. "If they know so much why can't they pay their employees?" Energia sent a fax stating they did not authorize these strangers to create the board. I learned informally that the fear was about some back door or corporate trap that would give Bermuda voting rights even after the new board was established, and so the Russians would be in the minority, and not the majority. Anderson again did the unpredictable. He announced he would give Energia 100% of the voting stock of MirCorp until the company was legally established. "This should end their paranoia" he angrily reasoned. "Why the hell not? They now control 60%, lets give them the whole stupid company and shut them up and move on to more important issues."

This was classic Anderson. His Russian partners were worried they would lose voting control, so he agreed that for now they could have the entire company, and he would have to trust them. But this was not a Hollywood movie, and the Russians saw the Anderson offer slightly differently. The radical move by Anderson set Vassiliev into a complete frenzy and put the management of Energia on full 'Defcon-1' alert.

Jacobson and I had an hour long meeting with Vassiliev and his aides.

We could not calm his worries about being given all of the stock in MirCorp. The next day I met with the Energia lawyer alone. It turns out Vassiliev suspected that a company having 100% of the stock of a Bermuda company was also 100% liable for the behavior of the company and Anderson was getting himself out of the legal responsibilities. Why else would Anderson give away the right to vote?

Making me as irritable as anyone was that I could only speak to Anderson when he called me, which was usually as early as midnight or sometimes as late as two in the morning Moscow time. I grew to loath my phone ringing once asleep, knowing it was Walt. Sometimes Walt had serious issues to discuss; sometimes it was with a simple question. My other problem with the late night conversations was given how far Energia was from the Hotel Savoy, Derechin insisted on picking us up by 8 in the morning. That night at around midnight I gingerly tried to explain this to Anderson. He didn't take the news nicely. "They are fuckin brilliant. Tell Derechin I'm gonna teach his people how to launch rockets."

And back and forth it went.

Enough is Enough

Other issues tugged at us so we moved on with the Bermuda issue unresolved. The Energia negotiation team tried to strike out the word "lease." This was the core philosophy behind Anderson's interest. John and I tried to understand the true reason for their sudden tactic. "Lease is not acceptable," said the lead negotiator, an older woman I had never worked with before. We 'tried' rent. That didn't work.

Energia was suddenly proposing that MirCorp should buy space station services. It was a subtle but critical difference. In our business model we were the operators, and in their suggestion MirCorp behaved as a customer. It seems Derechin or someone had decided that we would purchase certain scientific racks and certain rights to the cosmonauts. This was a major step backwards.

Jacobson was meanwhile struggling with the corporate resolutions. To our surprise Energia had written a complete set of resolutions for the board meeting. So too had Anderson, but I had no way to reach him. I again asked Anderson for his direct number. "Look," he briskly explained, "I'm involved in 15 companies around the world; I can't take phone calls from everyone all day and night."

We were heading straight for a major disaster.

I racked my brains to figure out why Energia was backing away from giving us 100% of the resources. For awhile we even thought that perhaps they didn't have the rights to the Mir, so we demanded and received a Russian copy of the decree of Prime Minister Primakov, from January of 1999.

Kristin Oland, my associate from the Energia Ltd. office, had arrived to help us with the board meeting. Kristin had other work in Moscow but knew, she told me, she would get sucked into the MirCorp work. She was right. Kristin worked into the night getting the decree translated. It stated that the Russian government maintained its ownership of the space station, but was transferring the full resources of the Mir to the private corporation RSC Energia for undertaking commercial services. Strangely, this decree had received little notice in Washington. Had I more time, it would have been worth publicizing this regulatory pinnacle of Semenov's long journey to be treated as a commercial company, and not a box under first Glavkosmos and later Rosaviakosmos.

Given the existence of the document, Energia's stepping back from the concept of leasing the station seemed a strategic move. This was one of the moments I had feared. Was some part of Energia, or the government, afraid to move forward with a plan as sweeping as envisioned by Anderson? My mole in the Kremlin provided little help, other than to repeat that the moderates were opposed to the deal and the communists and hardliners were supporting us, but only if we continued to respect their political realities. Deep Throatskii did add that allowing full, complete control of the resources of the space station was far beyond anything ever done in the space industry. Nor in any Russian industry. It would be like Exxon taking control of all the oil underneath Russia's soil. That's not a possibility, whether oil or gas or space.

The Energia negotiator finally suggested some split in resources, with MirCorp receiving commercial rights and they keeping the civil rights. We had to reject this suggestion, as we were afraid of leaving any 'marketing' door open. MirCorp needed to speak for the space station with one voice. Too often the Russians had killed themselves commercially by having one organization undercut the prices of another, or one spokesperson speaking at odds with another. What if we advertised five million dollars for a pharmaceutical research opportunity and the Russians, declaring their program "academic," advertised the same service for a million dollars? This was the awkward situation since the time of Glavkosmos and it was what Anderson and Tumlinson wanted desperately to avoid.

In vain John and I retreated into the bathroom for a private conversation. This despite the fact that Anderson had stated that we were partners, and no conversations should be secret from one another. I looked at Jacobson. "I've made no progress on payments or even what we are paying for, or how we determine the space station resources." Jacobson was having problems on the corporate resolutions. Energia remained fearful of taking control of the Bermuda company and hence little progress was being made.

It was time to show that we may be friends, we may soon be partners, but there were limits even to dreamers like Rick and Walt. There was no reason to move forward if we could not even agree on whether we were renters or customers. We returned to the small room outside of Derechin's office where we had sat for long hours every day. "We are leaving," I announced to the stunned negotiators. "Mr. Jacobson and I are going back to the hotel until day after tomorrow when Anderson arrives. We will not return until you agree to a lease of 100% of the station resources." Back to the Savoy we went.

Dinner that evening was a great tutorial on the Russian view. I invited a friend of mine studying in the railroad industry to join us. During a meal at Café Pushkin, the new three-story restaurant as sophisticated and comfortable as anywhere in Europe, Nellie Galtseva, a graduate student at the major railroad university, lectured Jacobson on the impossibility of foreigners ever taking control of the railroad lines. How could Russia allow foreigners, she wondered aloud, to control a strategic industry? This Russian student recounted the history of the railroads during the German invasion and the fear of the Russians towards surrendering vital industries to foreigners who may balk during times of national emergencies. John and I came away with a far better understanding of the huge hurdle we faced with Energia. It was no longer abstract.

After dinner Derechin called. He sounded hurt and disappointed. The conversation was somewhat comical. "You are fine?" he innocently inquired.

"Yes of course."

"I worry about your health, you left early."

"That is kind of you."

"Van at 8 o'clock tomorrow morning?"

"Do we have an agreement with the word 'lease' in the title?"

Pause. "We can discuss."

"No. I need an agreement with the word 'lease' in the title."

"You will not be disappointed."

What the Duma Wants the Duma Gets

The problems roared and smacked against John and I. It was numbing for both of us. We were confronting a revolving crew of negotiators each skilled in their own area. Energia wanted an agreement, but everyone up to Semenov was checking constantly to see how much authority could be retained.

The next problem was equally difficult to solve. Though concerned about the Duma's political response to leasing the space station resources, by mid-morning Derechin agreed to the language, and the lease of the space station by a foreign company was again the operating principal. An hour later he called me into his office. It was a standard Energia office, meaning bare walls except for the obligatory black and white photo of Sergei Korolev hanging behind the desk, and the latest Energia calendar tacked against the far wall. Sometimes there was a winter scene of the countryside on an Energia office wall, but that seems to be optional. Derechin's office did without the winter scene.

Slowly, carefully, he outlined his latest fear. It had to do with the Duma. The Russian Congress was behind the effort to shoot the film on the Mir. The Russian Duma may, in the future, want to send special cargo or people to the Station. What to do if a foreign company controlled all of the space station resources?

So that was the problem. First I tried business logic. "If they want to control the space station they should send the money they keep promising."

Derechin understood. But reality was reality. And he had another issue. Rosaviakosmos could not lose the rights to conduct civil space experiments from research organizations. Energia was afraid since we were a for-profit company that there would be no room for Russian civil projects, for which there were no funds.

I voiced my concerns. Suppose we actually found a wealthy person who wanted to fly to the station. We couldn't allow that person to be pre-empted by a Duma candidate. Also, who would invest in the company if the Russian government still controlled the station manifest? Another consideration was far more radioactive. Anderson hated all governments. He had made that point quite clear. I wasn't sure he would tolerate any informal understanding with the Russian government.

I proposed a two-part solution to Derechin. First, we would agree to fly Russian civil projects, though of course non-military, to the station at no cost as long as room was available. Secondly, we would allow the Russian government to go to the front of the line if they paid our standard commercial rates. Derechin was pleased but he had one more little problem. "The Duma has no money," he again reminded me. "If MirCorp has a foreigner willing to pay $20 million to visit as a tourist, the Duma will never pay you $20 million so their candidate will never fly."

So we were supposed to let the government fly for free?

The answer was yes.

I could never, never go to Walt with this proposal. I retreated back to Jacobson and Oland and we crafted a second proposal. That the Russian government could move to the front of our reservation line by paying our standard rate of profit to us. In other words, if MirCorp was making $2 million on a project that cost $20 million, for example, the government would pay MirCorp only the $2 million and Energia would absorb the service at no charge. I reasoned that Anderson would only care about the profit, and if Energia was absorbing the cost of the services, that was the same as being paid the full amount. Derechin checked with his senior management and the proposal was accepted.

Consider for a moment what was being accomplished, step by step. I assumed that Derechin had begun speaking with friends in the Duma, at Rosaviakosmos and in the Kremlin. All were quietly agreeing that MirCorp would control the resources of the Russian space station. Yes, the company would be majority owned by Energia. Yes, the Russians would control the board of directors, but the foreigners would make all business decisions and effectively control what research was being conducted on the station and by whom.

Based on the latest agreements, we moved forward writing the text of an agreement for leasing 100% of the space station resources, with the exception of civil science projects at no cost and a priority for the Russian government with payment from them at our standard profit margin. By the time of Anderson's arrival for the board meeting, a very tired John Jacobson and I could report that with the exception of the not-so-minor issue of Bermuda, the way was cleared for our formal ratification.

Chapter 24: At Least the Communists Are Happy

Reboosting the Mir to a stable orbit required raising the altitude at least 40 kilometers. There are two ways this can be accomplished. In the past, it meant use of the cargo ships. Boosting the space station 40 kilometers requires 1,100 kg of propellant, or about 2 valuable Progress flights. The Progress rockets were also being demanded by NASA for use supplying the ISS and also for boosting purposes. The Russian government has agreed to furnish the Progress ships for the next several years at no cost to NASA. The other option might be use of the novel tether concept, saving precious cargo ships.

The layers of mistrust between Energia and Anderson piled atop one another, while ironically at the same time all the required Russian players were coming together to support the novel agreement. The stumbling blocks were the Bermuda regulations, the secrecy of Gold & Appel and Semenov's delicate balancing act between appeasing the foreign investor

Investor Dr. Chirinjeev Kathuria at TsUP prior to the MirCorp launch. Kathuria brought a youthful enthusiasm to the project, as well as solid connections in media and financial circles. Looking on as always is the MirCorp problem solver for Russian bureaucracy Alla Botvinko.

and his domestic political supporters.

Because of Anderson's stubborn insistence to register MirCorp in Bermuda I had spent weeks trying to solve the unsolvable, and now found myself in early January visiting with the Bermuda lawyers. The blue waters and fair skies were a far cry from Moscow. Normally, this would be a pleasant location for business, but not on this trip.

On December 20th Anderson had sent a fax to Derechin expressing his shock and disappointment, saying that he was sure we would not be successful if the Russians continue to question each and every step in the formation of the company. "Rather than a season of celebration," Anderson had written, "we are falling further and further beyond. I want the resolutions signed by New Year's Eve or else I must re-consider this project." Never mind he had already sent millions of dollars.

At the core of the problem was disbelief by Derechin that the expressed motivations of Anderson were genuine. There was a chance, a risk, that there was a hidden agenda. As space engineers, any risk, however small, must be analyzed. Energia wanted to delay moving forward until the Bermuda regulations were studied to death. When Anderson snapped and gave them in writing a deadline for signing, it forced Semenov to move forward with the implications of Bermuda still not fully understood. The resolutions from Moscow arrived hours before the deadline, on December 31st signed by the Russian board of directors. Anderson would later sign personally since Bermuda never budged on allowing his venture capital firm to register. Nor did Anderson ever consider compromising and registering MirCorp elsewhere. I wouldn't understand just why until some years later.

True to form, whatever misgivings shown during negotiations, Semenov and all of the space company was prepared to fully implement the agreement now that an agreement had been reached. That meant that the moment had arrived for Yuri Pavlovich to battle against the policies of his own government.

Semenov needed to brief Yuri Koptev and receive his written support. It had to be done quickly and quietly. The timing was poor. Russia's holiday season starts with the New Year and then there is the Orthodox Christmas and finally Old New Year. The holiday's season really doesn't end until the 10th of January, so that would be the next critical date from Russia. While still in Bermuda, Derechin notified me that Koptev had been formally notified of the receipt of private funds for Mir operations.

At the hastily arranged meeting in Koptev's office there was yelling and screaming and anger and frustration shown by these long-time antagonists. Koptev was in disbelief. Surely Semenov didn't believe this unknown American would send the huge amount of money in time to change the Progress cargo manifest, now just two weeks from launch. Semenov then sprang the big surprise: the first payment had been received. The decree of Prime Minister Primakov was clear. Koptev's hands were quite tied. Ever the pragmatist, there must have also been some concern on his part that events were moving ahead quickly without his agency. If this was a patriotic deal, Koptev, being a government official, needed also to show the Duma he was not an obstacle.

The Energia lobbying campaign marched on.

Day's later Duma leaders were shown the proposed agreement, including the provisions for the majority control of the company by the Russians, and the commercial control of the station by the foreigners. The ultranationalist Liberal Party leader Vladimir Zhirinovsky, Socialist Head Vladimir Bryntsalov and reformers, including Boris Nemstov, the young liberal who had served in the Yeltsin government, and yes, Communist Party leader Gennady Zyuganov, all approved keeping the Station in orbit with the use of MirCorp's funds. My mole described the situation. Deep Throatskii, like everyone in Russia, was grappling with the shock of Yeltsin's dramatic resignation on New Year's Eve, allowing the ascension of the young Vladimir Putin as Russia's president. Presidential elections were scheduled for March. No one at Energia was certain of the full implications.

Deep Throatskii explained the political establishments' take on MirCorp as we sat in the lobby bar at the National Hotel. My friend liked meeting at the National, as it was far more normal for a Russian and a foreigner to be speaking English in a hotel lobby than one of the quiet restaurants he preferred. As he saw the situation, "first we are willing to take your friends money, why not? No one wants to sign the papers to bring down the space station. Only after taking the money do we worry about the political implications. This is Russia today. Take the money first and ask questions later.

"Surprise! Your intentions seem good. You personally are known to people here. And we have no other choice. The Communists and Socialists keep their Station. Fine. Let's move forward. Only Koptev is worried. His Americans are not happy. Relax, Jeffrey, he will figure out a solution for himself." From the Russian side it was all coming together. The Duma was fine. The first money had arrived. Putin's cabinet was meeting next Thursday to formally approve the deal. Koptev understood he must now notify Dan Goldin. In the quiet of the Russian Christmas season the letter was drafted and prepared for Koptev's signature. It would be sent once the cabinet approved. Far away, in Bermuda, I held my breath.

First Impressions of MirCorp

We would live and die on how MirCorp was first introduced to the world media. It had to be done right. There was every chance the project would be branded as a bunch of rebels working with the commies against good old NASA. That perception would make a difficult project even more difficult. I was working on the assumption that the first round of funding from Walt would be there. And that the Russians would honor the agreement. I felt the tether would work and so much of 2000 would be spent on finding investors outside of Anderson. The Company had to be introduced in a professional manner in order to attract serious funding.

Anderson was pushing me to work solely on the organizational issues, but I took the time to bring in a public relations expert to begin crafting the media message. Jeffrey Lenorovitz was a former reporter who came from years of working as the Paris correspondent with the industry magazine, *Aviation Week and Space Technology*. He now ran a public relations office near Washington, DC with offices in Paris and equally important, in Moscow. His clients were a number of European companies, including Arianespace, the huge French launch vehicle effort. For many international reporters and officials, Lenorovitz was an 'impartial' American, one who spoke French, had a French wife and had worked for years with colleagues from different countries. That was critical for MirCorp. We could not be viewed as an American effort, or have a spokesperson who had worked for a major aerospace contractor like Boeing, or the Duma will cut our heads off.

Lenorovitz was always a breath of fresh air. His tone was cheery, very French-like. "Jeff here," he always called out on the telephone. "Got just a few minutes?" He understood very well the complex task we faced. It was not a single story. It was a story of working with Russia; a story of transforming NASA; a story of exploring space in a way different than any other manned space company. "Reporters today have so little time," he explained. "We have e-mail and web and phone calls and faxes and there is a tendency to simply rewrite the press release. That won't really be possible with MirCorp. We want them to dig a little deeper, understand just why we are doing this."

Tumlinson was disappointed in my choice. He wanted MirCorp to retain a more prominent PR firm. That could come later, and I told Jeff that perhaps his assignment would be temporary.

Just before leaving for Bermuda we produced the first working paper on our message. MirCorp would be positive at all times. Never critical of NASA. We would stress our international structure, our peaceful operation of a space station without government funding. That the Mir was a symbol of a new generation in the post-Cold War world. That our values reflected that of young people everywhere: no military, no national boundaries. Space exploration for the pure pleasure of space exploration. "I don't want to be involved in something that is negative," stressed Lenorovitz. "We can win if we are positive."

We also agreed to de-emphasize Anderson. MirCorp was not a one person venture. It could not be. If it was presented as an Anderson project then it could be dismissed by NASA as a grudge move by a wealthy investor disgruntled with American space policy. I wanted to stress that the funds were flowing from the venture capital arm Gold & Appel, with a second investor, Dr. Kathuria already signing on. In short, it was a sophisticated message that required everyone to sign onboard. Walt understood completely while Rick was a nagging problem. His goal in keeping the Mir in orbit was to send a message, or more like a grenade, to the supporters of the status quo. Not building a company.

NASA would be pissed off enough. Boeing would be furious and threatened enough. We didn't need someone associated with us to rattle the cages of the government and the dinosaur space companies. One of my biggest challenges was having Rick understand that freed from being part of MirCorp he could say and do what he wanted. As part of us, it was a different story. While in Bermuda he had already sent me an e-mail saying "I am having post-birth trauma at having to let go of MirCorp. But I know it's in good hands with you." That sounded ominous. Lenorovitz thought so as well. But I was pretty sure Rick would focus on the tether project and let us develop MirCorp into a serious commercial venture. I promised Rick that when the time was right we would put him on the stage and "let me have a go."

Lenorovitz and I decided that the best shot at creating our image was a press conference for the signing of the lease between Energia and MirCorp.

He wanted it in America, as did Tumlinson. I wanted Europe and we settled on London. Derechin agreed. Semenov and I would sign the lease giving MirCorp the resources of the space station at a public ceremony. To my surprise, Semenov agreed to travel to London, along with Victor Legostaev to not only sign the lease, but take questions from the press.

The Leak that Scuttles the Plans

It was not the leak on the Mir space station that caused so many problems in January of 2000. That leak was understandable and, we all hoped, could be fixed if a crew could be sent to re-open the dormant space station. No, it was the leak to the media that hurt us so badly.

Sergei Gromov had to be acting under orders from someone. There is no other explanation for what took place on Monday, the 10th of January. Gromov was one of the liaisons between the media-shy Energia and the Russian press. His job was usually to keep away the prying eyes of reporters. That Monday afternoon Gromov brought some Russian reporters together and informed then that an American company named Gold and Apple (sec) had provided $7 million for continuation of the space station.

This was how Dan Goldin and NASA found out about the project.

Why did he do it? Gromov always denied he was the source of the leak, but there was little doubt of his involvement. My feeling was that Energia had to force the hands of someone who was reluctant to move forward. Maybe Koptev. And the best way to do so was to announce that the funds had already been received. On the 13th of January Valery Ryumin went on Ekho Radio, a large station in Russia, and confirmed that the funds had been received. Someone must have been resisting, else this shy company would not have taken these steps.

Rick Tumlinson considered the leak a betrayal of the collective promise to keep our mouths shut. "What the hell is going on? I've been a goddamn angel and now your Russians are leaking. Can't we trust them?" Rick got hold of me on my last day in Bermuda. He was so distraught he was breathing hard. But his anger was not for the reason I was angry; that we stood to lose control of the media situation. Nor that Gromov had called us an American venture, which was the very identification I sought to avoid. Or that he had given the wrong name for the fund or announced it before

Goldin knew. No, it was because Tumlinson wanted to tell his story to the press. That would still be possible, if Rick showed some patience.

Anderson was also furious. In the morning before heading to the airport leaving Bermuda I sent off an urgent fax via Kristen Oland to Derechin, asking whether "Gromov worked for the devil or for NASA."

The situation went from bad to worse.

All the preparations on media strategy were hemorrhaging away. Seth Borenstein, the Washington bureau reporter for the Knight Ridder papers, a chain of mid-sized dailies throughout America, telephoned to say he was going ahead with an article that the Mir space station had been saved by Western investors. Interestingly, he didn't seem to know about the stories in the Russian press. This was Tuesday the 12th of January. Goldin had still not been informed and so I refused to speak to Borenstein. He called again. I knew Seth and knew he was a professional. "Tumlinson gave me the whole story," he immediately let me know.

I began sweating. Then the real bombshell. "Is this guy Anderson really a recluse, like a Howard Hughes?"

"Who the hell told you that?"

"Tumlinson."

It was my turn to get to Tumlinson.

"We agreed." I was screaming. "I don't want the first story to be a regional newspaper chain. We agreed it would be in Europe. We agreed you would remain quiet. And a recluse? God, Rick, if he makes that the slant we are dead. You understand? That's why we had an agreement, Rick."

"That was before your Russian friends spilled the whole situation. We've got to get our story out." And with that the conversation was over.

Having no choice I worked with the reporter to get some good quotes in about our business model and the investment potential, anything to try and deflect it away from the focus on the nutty investor. I got hold of Anderson and told him to order Tumlinson to shut up and by the way, your good friend is calling you a recluse, which could well screw up any chance we have to get investors. Anderson couldn't believe Tumlinson would say that. I had no time to debate any Freudian issues between the two guys. Lenorovitz and I scrambled to find some saving grace. "I don't want to be a smart ass, but I've been saying we've got to get Tumlinson to scale back his anger," psycho-analyzed Lenorovitz. Easy enough to say, though he was right.

Still needed was a damn European article and fast. We had to have some global slant to this announcement. I offered Tumlinson a carrot with the stick. "You pick a British reporter and I'll cooperate. But do it quickly," I pleaded.

The first U.S. story came out on Thursday, the 13th.

U.S. tycoon wants Russian space station Mir to be a vacation spot

"A reclusive American telecommunications tycoon is spending millions of dollars to renovate Mir, the closed Russian space station, into a commercial lab, advertising gimmicks, and out of the this world vacation spot for his fellow millionaires."

Reclusive. Gimmick. Vacation spot. Fellow millionaires. Terrific.

It could not have been a worse introduction in America. Borenstein drove the stake in our heart by writing at the end that "Tumlinson, who hooked up Anderson with Energiya, calls the financier, "reclusive." Borenstein had the story right and it was even-handed. He quoted a Russian analyst as saying this deal was "feasible." He had me explaining our emerging business model. The reporter also quoted two of the NASA astronauts who had lived on the space station. Jerry Linenger was critical. But John Blaha, the third astronaut to visit, stated he "would tell anybody it would be the most significant thing they would ever do in their lives."

All the preparatory work with Jeffrey Lenorovitz may have been for naught. It was a terribly frustrating moment and confirmed my fear that events would always be a step ahead of us in this project. Too many people had different motivations in saving the space station; especially given that we had the receptive ear of the world's media. That was an intoxicating mix for many, whether they lived in California or Moscow or points in-between.

On the very same morning that the article broke, the letter was finally transmitted from Koptev to NASA's Dan Goldin.

(Edited Copy)

13.01.2000

Mr. Goldin
NASA Administrator
Washington, DC

Dear Mr. Goldin:

In compliance with the Decree of the government of the Russian Federation that allowed the RSC Energia to continue operating Mir after the first six months of 1999 using off-budget sources of funding, the Russian Aviation and Space Agency considered the RSC Energia proposal on the Mir Space station program.

Under this program, the operation of the Mir Space Station is going to be funded from off-budget sources and the station will continue its operation until August 2000.

Under this program, we plan two cargo missions and one manned flight to the Mir space station.

The RSC Energia has succeeded in attracting both foreign and domestic investors who can provide funding for the proposed program. We also plan to use a portion of these off-budget funds for the ISS program. In light of public opinion and the political situation involving the Mir Space Station, I would like to tell you the essence of the decisions that we have taken, so that you have a better understanding of the proposed Mir program that is due to be approved by the Government of the Russian Federation this January.

Once again I expect you to fully understand the situation and count on your support.

Respectfully yours,

Yu. Koptev (Signature)

General Director

The letter was a thunderbolt that struck NASA. Without official warning it was a done deal. A betrayal of the agreement between NASA and Rosaviakosmos. Between the Russian Federation and the United States. Most personally, it was a betrayal of the friendship between Dan Goldin and his good friend Yuri Koptev. "Shocked and disappointed," were the words the Administrator used to friends and colleagues over and over. Goldin spent several days in a serious funk. One of his first questions to Koptev, according to colleagues of mine in Washington, was to ask why he wasn't given any warning. Why he had to learn of the decision from press accounts out of Russia. Koptev was being very careful. He waited until the Russian Federation had made the official decision and then notified his friend at NASA in writing and then with a phone call. Also lost on NASA was that the letter arrived immediately after the end of the Russian holiday season. No time was wasted. And it couldn't. We had so little time.

NASA would not have been surprised had Goldin taken the time to understand Russia. For years he had allowed his personal animosity towards Semenov and Energia to color not only his behavior, but to help shape American policy towards Russia in a way harmful to the International Space Station, and harmful to relations between NASA and Russia. To Goldin, the only hope for good relations was his friend "Yuri." Semenov and Energia were to Goldin some sort of evil force keeping Russia and America from working well together in space. To Goldin and the American government, Semenov's agreement with MirCorp was irresponsible. NASA and the contractors had been told throughout the last year that "Russia" had ceased its funding of the Mir space station. That "Russia" would throw its support behind the International Space Station. But since Prime Minister Primakov had transferred the operations of the Mir to Energia, Koptev spoke only for the federal budget and his agency. These sorts of nuances were lost on Goldin and NASA and State Department and many in Congress.

There is some mystery to the letter. It refers to 'domestic' investors. It is an important point. Lost in all the uproar was the fact that Energia was also commercially investing its own resources in this effort. Our promised $20 million would not have gotten us very far, not even with Energia. Energia was investing an equal amount in services. I'm not sure what Koptev meant by also investing in the ISS. That might have been an olive branch thrown to the United States to help his friend Goldin sell the changing situation to an angry Congress. Or an understanding that with business from MirCorp, Semenov could continue the absurd practice of using commercial funds to pay for government obligations.

By the next day Tumlinson was working with British reporter Tim Radford, the science editor of the *Guardian*. Radford had followed the tether project and was rewarded by Rick with the story. Again, it was filled with such *bon mots* from Rick as "There is a witch hunt going on here, about tech transfer, even though the air force people say this (the tether) is not a threatening technology.

"Nasa hates this," says Mr Tumlinson. "Imagine you are Nasa, you have spent all these years conning the world into believing that your $40bn space station is going to solve all these problems, and somebody opens up a little shop down the street. That's what's happening here."

I stuck to the script. "It will be an international company, located in Europe, backed by people worldwide, and we are not buying Mir. Mir is the property of the Russian government. We are creating a partnership with Energia that will together market and raise funds and enter into corporate partnerships that are appropriate," said Mr Manber. "That's about where we are now. We are expecting to sign with Energia and hold a press conference in a couple of weeks."

Only one news service had a different slant. That was the French news wire AFP. It's worth mentioning, as this was the first news service to focus on the true significance of our venture. Not space tourism or whether Anderson was a nut, but the extraordinary change in the behavior of the former Soviet space program.

Capitalist Revolution to give Soviet-era Mir a space lift

Moscow, Jan 20 (AFP)

Russia is teaming up with Western investors to give the aging Mir space station a new lease of life, with the dream of bringing corporate sponsorship, the Internet and tourism into the space age.

Mir's operator, Russia's Energiya corporation, has taken a majority stake in the newly-created MirCorp set up to attract investors to the project, which will run Soviet-built Mir along resolutely capitalist lines.

"Here is a project started by the Soviet government, and now it's a joint international project," MirCorp President Jeffrey Manber told AFP.

Reporters in Europe were more willing to look at the project based on its merits, and its historical significance not just for the industry but for Russia as a whole. Those writing in America were concerned with how we were a harmful threat to the NASA space station. In Europe, they appreciated the historic nature of MirCorp. And, no matter how great the commercial challenge, a major hurdle had already been overcome. With the signing of the lease Russia would be making a historic step towards the commercialization of its space program. Typical of the "us vs. them" story was the piece by Mark Carreau in the *Houston Chronicle*, which ran in part:

The Russians are responsible for furnishing the Soyuz and Progress crew and cargo capsules that must be regularly launched to the new station. A renovated Mir would require a similar number of launchings, possibly competing with the new station for production of rockets and capsules.

This week, NASA said it will embark on a fresh assessment of those Russian production capabilities. It plans to complete the audit before experts from the two countries meet next month...

Why are the Russians, the cash-strapped Russians, the bankrupt Russians, "responsible for furnishing" the capsules to the new NASA station? Why isn't NASA paying for these services? This is the policy question that should have haunted both Dan Goldin and Yuri Koptev. How could Goldin rest the future of NASA's International Space Station on launches provided by the bankrupt Russian government while ignoring for years the commercial path offered by Semenov?

We decided there had been enough publicity, and Derechin, Anderson, everyone agreed to refrain from public comments until the press conference. Anderson got Tumlinson to agree as well. I should say it was not Tumlinson's intent to focus the story on Anderson's social habits. He did mention our scientific interests and the economic viability of the project, but he was unaware, or caught off guard, by how cut-throat are the major reporters. Small talk both before and after the formal interview is often where the most controversial comments in a story come from.

In Russia everyone was talking about this mysterious source of money called Gold Apple. Anderson later gave me a gold-colored brass apple to commemorate the moment when our battle truly began. A few months later I would discover the significance to Walt of the Golden Apple.

Chapter 25: The Mir is Saved

On February 1st the Progress cargo ship known as M1-1 roared to life from the launch pad at Baikonur. The rocket carried the expected several tons of fuel, but it also contained food supplies and 150 kilograms of compressed air. The mission of this cargo ship was unprecedented. It had been developed by a private Russian company assisted with funds from a privately held international company. The mission was to boost the space station to a stable orbit in order to, frankly, figure out what to do next.

The compressed air being carried by the Progress would be used to make up for the loss of atmosphere, resulting from a slow leak afflicting the station. TsUP was reporting that the atmospheric pressure was now down to 580 millimeters. The lowest acceptable level of air pressure for the crew was 560 millimeters, which would certainly be the situation in another month if no action was taken. The plan was to remotely pump the com-

NASA officials were stunned how each major area of expertise had in Russia a noted leader. In America officials are rotated in and out of programs and consequently personal responsibility is minimized. (From left to right). Oleg Lebedev of Energia was the specialist for the earth observation systems on Mir. Victor Blagov was the deputy director at Mission Control for decades and understood fully all communication issues between space stations and earth. Leonid Gorshkov was the senior specialist who during the initial NASA meetings realized the plans for space station Freedom were flawed. After extensive internal deliberations the drawbacks were discussed first with Boeing and then later with NASA. Photo: Energia Ltd

pressed air into the station. If that failed for some reason, then the next crew would have to enter wearing spacesuits and do the job themselves.

It was a technically demanding task to keep the frozen and leaking station in a healthy state of hibernation. In early January the ground controllers at Korolev had put the Mir into a slow spin, allowing it to sort of bask in the sun, providing full exposure for the solar panels. The decision was then made to delay the repressurization of the station until just before the crew arrived. Until then, the Mir would stay asleep, with the main computer in shut down mode.

We now expected the MirCorp manned mission to take place in late March or early April. But an announcement could only come once more funds were sent, and the Russian side formally agreed to support another crew to what had been thought to be a permanently dead outpost.

Jeffrey Lenorovitz busily worked the media, helping them understand the uniqueness of the launch and its impact on keeping alive the space station. We received good coverage, along with the growing backlash from NASA supported media outlets.

As hard as Jeff worked with reporters, NASA worked even harder. It was not just the Progress that had roared to life; so too the counterattack from NASA. Stories came out that we had taken from NASA their cargo ship and somehow MirCorp, Energia and by extension the Russian government was tricking NASA. MirCorp was portrayed by some as anti-American and negative. Lenorovitz worked with reporters from the *BBC, Financial Times, New Scientist, NOVA* as well as those in Asia, and those U.S. reporters willing to listen, to show us as pro-space, pro-free markets and supported by Western values. Oh, and we didn't steal NASA's rockets.

Now that step one had been accomplished, my attention was focused on building the company. Walt was insisting that the office be in Holland, as this was the location of his greatest commercial success, a telecommunications company he built from scratch. It also made some limited sense from a space perspective, as the Dutch coastal town of Noordwijk was home to the European Space Agency's space station program. Walt insisted we not hide and to the contrary, should locate right alongside all the European space station contractors. The problem with this line of reasoning is that it assumed our business would be traditional space station research and development. It seemed far more likely our customers would come from outside the industry. London might be better suited in that regard, as it was home to the European financial markets, media and growing Internet platforms.

MirCorp BV, a Dutch company was created as a subsidiary of the Bermuda company. Never mind that the Russians were still balking at taking the stock from Bermuda out of fear of the unknown. With the Dutch company formed, we could now finally get an office, complete with landlines, good computers and all the other nice things taken for granted.

Until the office was located we had to task and trust different people scattered about to implement what needed to be done. Critical was a tall, good looking Belgium named Gert Weyers. A friend of Kristin Oland's, Gert was soon indispensable, as in addition to fluent English, Gert spoke Russian, French and a few other languages. He worked for us first in Holland, and then later in Moscow. We also relied heavily on Andrew Eddy, the suave junior official from the Canadian Space Agency in Montreal. Andrew had the practical experience of working every day with those government officials focused on the International Space Station, and was hired to figure out the scientific and research part of the puzzle. We had two senior executives who ultimately didn't last, and Kathuria stepped in as our interim CFO for the critical first months.

Right off the bat the need for new investors and the need for customers seemed to be of equal importance. Both of our current investors were confident that their network of colleagues would be attracted to MirCorp, but what was needed was the right sort of package. Kathuria pulled together a strategy with Eddy that involved the writing of our business plan, backed by a valuation from a respected accounting firm regarding the asset of our lease, the Mir space station. Kathuria's idea was that if an independent source bought into the idea of Mir as an asset, our valuation would be in the hundreds of millions of dollars. Seeking a hundred million in investment capital against the asset would therefore not be as rash a strategy, even given the political uncertainties. Eddy had worked with the Montreal office of KPMG, and we hired them to undertake an independent analysis of the value of the station, as well as to work with us on the writing of the business plan.

For the business plan we turned to the well known McKinsey and Co, which pulled together a two hundred page working document that explained how the space station could be turned into a commercial platform. For long term strategy Kathuria had connections to senior people at a major international consulting company located in Chicago. After several meetings, and some prompting from the Kathuria family, the consulting company agreed to undertake a comprehensive study with recommendations on companies that would be interested in using the space station for science, for branding and for media. Critical to me, and I hammered the point home in one Chicago meeting, was that not only would a report be produced, but the firm would also introduce us to the identified companies. The cost for their services was not inexpensive, but it would be worth the price if introductions could be made across U.S. industry by September.

We had somehow managed to accomplish both the technical starting points and the commercial. The Mir has been saved. A successful Progress launch gave us the time to design a business strategy that could, with a bit of luck, allow us to implement our business model. Unknown was just how would NASA react.

Chapter 26: Some Get It, Some Don't

Use of an electrodynamics space tether, if successful, would dramatically reduce the need for conventional rocket thrusters for boosting purposes on both the Mir and also the International Space Station. A long thin wire would be attached to the Mir, which would function as an electric motor, exerting a small upward force on the Mir as it travels through earth's plasma field. After ten months of use, the Mir would be re-boosted the required 40 kilometers. Thus, use of the tether would open up more Russian hardware for the ISS, potentially save the cost of using Progress ships for the ISS, and providing NASA with a new technology for ISS energy requirements.

Presentation given to the U.S. Department of State

December 15th, 1999

Seeking Export License of Tether to Russia

Two key players with MirCorp were Jeffrey Lenorovitz (left) who was a veteran industry reporter who covered extensively the Soviet space program. He crafted a media message that stressed MirCorp as a symbol of the post-Cold War era. Andrew Eddy (right) left the Canadian Space Agency to handle our marketing to ISS partners. His experience fulfilled Anderson's desire to compete against the International Space Station for commercial industrial space research.

Photos: Jeffrey Manber

A critical "day" in the life of MirCorp began in Washington, DC on February 16th and ended on February 17th in London. Two events, separated by so much distance, were inextricably linked-as Dan Goldin reacted in a Congressional hearing with frustration and anger and misunderstanding to the MirCorp project.

The London event was the formal signing of the lease between MirCorp and RSC Energia. The signing took place in the Institute of Directors Club, located on Pall Mall, an august street just down from Piccadilly Circus. Present from MirCorp was Yuri Semenov, Victor Legostaev, myself, Andrew Eddy, who we had recruited from the Canadian Space Agency and the new investor Chirinjeev Kathuria.

Dr. Chirinjeev Kathuria was a living poster boy for the Internet business model. An Indian-American Sikh, he looked striking in his red-turban and wide, wide deer-like eyes. Though young at 35, he had already amassed a fortune by creating a number of free portal and other Internet ventures worldwide. President and co-founder of the New World Telecom, a founding Partner of the X-Stream Network, Co-founder of Live Door out of Japan and involved in telecommunication projects from Nigeria to Albania to India, he had a personal fortune of several hundred million dollars.

Anderson and Kathuria met while working on a telecommunication project and the two soon discovered they shared a passion and dream about manned space exploration. As a child Chirinjeev harbored hopes of becoming an astronaut and pursued a medical degree with the thought that one day he would work in space. Like so many of us, the slow pace of NASA soon dashed those dreams. In the lobby of the Club Chirinjeev was bubbling with excitement about being in London and being part of the project. The young Sikh in the striking red turban drifted around the lobby, coming in and out of our conversations.

Anderson had declined to attend; more and more the recluse label seemed to have some merit. Legostaev kept asking why the major investor couldn't make it to London and I really had nothing to say.

Packed with print and television media, the room for our press conference was vintage British, with huge oil paintings of forgotten business elders and 19th century war heroes. The Russian correspondent from the powerful Channel One Television came up to me as soon as I walked in; expressing amazement that Semenov was holding a press conference in a foreign country, something he was loath to do. The reporter was right. At one major space event in the United States Semenov had refused to speak to the reporters, and when one tried to push his way into our circle, Semenov simply turned his back. The young reporter was appalled. "Doesn't he understand the power of the press?"

Down the hall in the Institute of Directors Club was another press conference. This media event was for the software giant Cisco Systems, and attracted maybe ten reporters compared to our forty or fifty. This is not a boast by any means. Cisco was announcing a new software product worth hundreds of millions of dollars, and present was only a few bored industry reporters. As a business MirCorp was a dicey proposition, yet it was again testimony not to us but to that special place that the space frontier holds in the imagination. Just as with Payload Systems, just as with Richard Branson's Virgin Galactic today, the attention is not due to the bottom line but the excitement of the conquest of space.

The announcement went fine. Semenov was in a good mood during the one hour event and answered all the questions fully. The only rough spot came when I announced that as part of our business plan, we expected to send "citizen explorers" to the Mir space station, at a cost of some twenty million dollars per ticket. Citizen explorer was a term invented by Tumlinson, as he felt strongly that space was not yet for tourists. The reporters laughed. The thought of someone spending millions to ride into space seemed absurd to them. Never mind that a British woman and a Japanese tourist had already made the journey.

After the event Semenov and Legostaev of course went their own way. Chirinjeev, Kristin Oland, Jeffrey Lenorovitz, Andrew Eddy and myself went out to a Russian bar in central London and celebrated with vodka and dishes named after key events in Russian history.

Back at the hotel, amidst the congratulations from colleagues and strangers were a dozen unexpected messages, alerting me to that first event that had taken place the day before in Washington, DC. NASA administrator Dan Goldin had testified before Congress the day before our lease signing ceremony and had slammed both RSC Energia and MirCorp. Goldin unleashed his frustration at an event known as a 'posture' hearing. It took place before the Subcommittee on Space and Aeronautics in the House of Representatives. The chairman was Dana Rohrabacher, the colorful congressman whose previous credits include professional surfing, a vote-getting asset considering his district was in Southern California. Rohrabacher was a true believer in removing NASA from space station operations. He therefore had an intuitive feel that the commercialization in Russia was the right path and should be supported more by the United

States. The California Congressman had expressed his philosophy as early as 1995. In a speech before an industry group Rohrabacher voiced his belief that the space station should serve as "a hub in space for private activities, instead of a bureaucratic stronghold in space for government. I do not see it as a fort there with cavalry inside. I see the space station as being a trading post where you have people building different things around the space station, and you have an interaction with the private sectors. I see no reason why we can't have private companies supplying space station. I don't see why we can't have man-tended industrial platforms around the space station that can be checked on, with a commercial contract by people in space station."

He and other Congressmen, including Robert Walker, a far-sighted Congressman whose district included the Amish of Pennsylvania, had long been troubled that Goldin refused to allow NASA funds to be sent directly to the Russian industry actually manufacturing the hardware, and instead had pinned NASA hopes for Russian cooperation squarely on the shoulders of the small Russian Space Agency. Years before I had spoken with Rohrabacher's chief space staffer, James Muncy. A dedicated warrior for commercial space, Muncy had called me sometime in 1995 when Goldin was ratcheting up his criticism of Semenov and Energia. In the course of a long phone conversation I drew the distinction between Semenov as a tough negotiator and a Russian patriot, rather than someone somehow against American values. I revealed my ownership of Energia stock; foreign organizations also owned stock in Energia, and wasn't that what we all wanted from Russia? Certainly Congress could understand this. Muncy was shocked at how far privatization had advanced within the Russian space program. He asked for some visual prop that would be easily understood. So I made a photocopy of the Energia stock certificate, covering up the number of shares, and sent it over to Muncy. I'm told Rohrabacher held up the certificate on the floor of the Congress to illustrate the value of working directly with the Russian space industry. A stock certificate from the once-secret military Soviet organization that was now a Russian company cut through all the endless debates on the value of working with the Russian manufacturers.

At the hearing Rohrabacher confronted the NASA administrator in some pretty blunt language. As Goldin listened from his witness table, Rohrabacher proclaimed how "nothing has been so destructive as the naive assumption that Russia's government would spend its limited resources to help build the International Space Station.

"While commercial relationships with America launch companies motivated the Khrunichev and Energia corporations there in Russia to clean up their acts, the Russian Aviation and Space Agency oversees many entities that sell arms and technology to rogue states.

"The sad thing is that there was another path that could have been taken. For the last 3 years, Chairman Sensenbrenner and I have been pushing the Administration to change this form of dysfunctional partnership with the former socialists and the people now in charge in Russia to a market based approach, which would have had us having direct relationships with Russian corporations in strengthening the private sector approach, and the private sector in Russia." The Congressman continued his own walk through the recent history of Russia-American space cooperation, as seen through the eyes of NASA.

"As recently as 2 years ago, we heard NASA witnesses tell us that they couldn't simply sign contracts to buy Russia's critical path contributions to the Space Station because if we just gave the money to (the) private sector corporations, Russia's' pride would be hurt.

"Then NASA sent the Russian Aviation and Space Agency $60 million so they could meet their commitments. In fact, as far as I can tell, the only reason for continuing the pretense of this socialist partnership with Russian bureaucracy is the pride of a few people on this side of the Atlantic, not on their side."

Arnie Aldrich and Sam Keller, and many others who had understood the correct path was commercial relations with the emerging commercial Russian space industry, must have been ruefully smiling. The many people who advised Goldin to overcome his personal objections to Semenov must have also been appreciative at the honest words from the Chairman. The hearing moved on but Rohrabacher later returned to his theme about commercialization, asking Goldin to confirm that the Russian companies have a pretty good track record in terms of commitments, whereas the Russian government does not.

Goldin's answer ignored the question ´´Let me provide some perspective,`` the Administrator began. "Let us not forget what we learned from the Russian government in the Shuttle-Mir series of flights. It was monumental in opening our eyes to understand the rigors of long-term space flight. And that occurred not because of the Russian companies, but that occurred because of the relationship at the top level with the Russian government.

"And Russia," continued the Administrator, "at the time and for the last number of years is not capitalistic in nature, and the Russian companies do have to work with the Russian government." Good thing I wasn't in the room. I might have jumped up all crazed at hearing this nonsense and been led away by Capitol security.

The real issue is that the Russian space community would have become more capitalistic in nature if it had not been suffocated by one of its largest potential customers, the United States space agency. Boeing was working with Energia on Sea Launch; Lockheed with Energia and Khrunichev on the Proton launch vehicle, to name just two huge examples. In these instances American industry benefited economically and technically from working with the Russian organizations on a commercial basis and not through the chaotic Yeltsin government. And the projects came in on time, on budget, no more and no less than any other large aerospace project.

The price tag for ISS would have been higher in the beginning had NASA not forced the weakened Russian government to provide the launches and a core module at no cost. Still cheaper and safer than if relying on American industry to manufacture a totally new cargo system and habitable module, but higher none the less. The Goldin path led to several onerous results. These included an unfair black-eye for Russian industry, a long-delayed space station as the Russian government could not honor its promises, and ultimately far higher station costs. Not to mention assuring that commercialization never took off within the Russian industry.

Goldin then vented his frustration over us. "Now," said Goldin, "let's talk about some Russian companies. There is one Russian company who quoted us $65 million for a Soyuz vehicle. And they turned around and sold for $20 million 45 days operation on the Mir space station, one Soyuz and two Progress vehicles."

Wryly commented Rohrabacher, "So we didn't get a very good deal?"

The Congressman continued on this theme of who bore the responsibility for the late space station modules by asking Goldin that "if you make an agreement knowing your partner cannot live up to it, aren't you as much in fault as your partner?

"Someone told me their total budget is $35 billion for their whole government. We cannot expect a country with that small a budget to transfer resources over to a space program, even if they made agreements with us. We're fools for depending on them."

But the Administrator was a fighter. In answering a question from Congressman Weldon, whose district includes the Kennedy Space Center in Florida, the administrator admitted that "I'm shocked, upset, disappointed, I don't know how many more words I can say, about the fact that I mentioned before, for $20 million a US company has negotiated an arrangement with a Russian company, who is one who works with us, to have two Progress vehicles, one Soyuz vehicle and 45 days of operation. We are not happy. For that reason, we withdraw any requests to this Committee to buy a Soyuz vehicle for $65 million. We're going to take a good hard look at this "international" pricing and internal pricing that the Russians have before we commit to another nickel."

No understanding of market capitalism was displayed by Dan Goldin. No understanding of the difference between a contractor price, such as that charged by Boeing using government money, and that of commercialization, where two partners are both contributing resources.

Several long hours later the Chairman returned to his opening theme, the decision by Goldin to avoid working with the Russian space industry, and to rely on Rosaviakosmos to provide rockets and hardware at no cost. Reading about the hearing fresh from the London press conference, I wanted to believe that the frankness on the part of everyone was partly due to their surprise at the ramifications of MirCorp, which must may have shook up the normally quiet Congressional hearings.

"Mr. Goldin, just one last note here. Now, first of all, didn't the price that Energia charge us originally, didn't that come through Rosaviakosmos?"

Goldin again sidestepped the intent of the question, stating they had a team in Moscow last year and had traced the American money and that the Russian space agency was not holding anything back. This was some funds sent in exchange for the Russians giving up a large portion of their share of their own module, the Service Module. My belief was this move was also a NASA mistake, as it reduced the role of Russian researchers in ISS and made them more likely to cling to the Mir.

"Well, I don't think that we're talking about holding back money" correctly retorted Rohrabacher. "I think that this indicates to us that having a private sector company move forward in trying to establish this relationship has given us a better understanding of what the market price for Russian hardware and technology is."

"But this is the Energia company," said the pleading head of the U.S.

space agency. "This Administrator (fascinating his sudden use of the third person, as if trying to pretend the opinion was removed from his personal ego) is not pleased with the performance and the attitudes at the Energia company. And its-I understand the direction you're going in, but I have to deal day in and day out, it is the director of the Energia company who pressed real hard to keep the Mir station up. It is the director of the Energia company who proudly walked me through his plant, and identified the tail numbers and thanks us for the support we gave so they could build them.

"It is this same person who, without any consultation with NASA, pulled those tail numbers to keep the Mir space station up. I'm speaking with intensity and emotion. And this Administrator has a great deal of concern of dealing directly with the Energia company for the reasons I have just stated. However, it is not that we don't have tools, and it is not that we won't have fervent, frank and candid discussions on this pricing equity."

Was the mention of having "tools" a threat being voiced by Dan Goldin?

Every once in a while a complex human being bares his or her soul, and that's what Dan Goldin did that day in Washington and later with a reporter from *The New York Times*. During the *Times* interview, in his Washington office, the Administrator admitted, in what the reporter describes as his thick Brooklyn accent, "My feelings are bruised. I have a hurt. I'm not saying MirCorp shouldn't have done what they did. I'm just saying I'm in the book, they've got my number and it might have been nice if they called."

Goldin raised with the *Times* the issue of the funding to the Russians. "I'm not saying that money is being diverted," Goldin protested, thereby intimating that the Russians may be siphoning money from the International Space Station and dumping them into Mir. "I'm just saying the arithmetic doesn't add up. So like I said, I have some bruised feelings. And the arithmetic! I'm confused."

For the next few weeks I was forced to answer to reporters and government officials just why MirCorp received a far lower price than NASA. This included even a meeting with the scientific advisor to Vice President Gore.

Let me give you my explanation, though I'm sure you, the reader, understand how market capitalism works. Our cash payment to Energia covered most of their basic out of pocket cost that had to be paid to their own vendors. In addition, Energia is MirCorp's contractor for Mir space station services. If a space tourist flies to the station, a large portion of the funds go to Energia. Say the ticket was for $20 million, then Energia would receive another $18 million.

But Energia is not only a contractor to MirCorp, it is also a partner, and owns stock in the company. Should MirCorp be commercially successful the shares will appreciate, and Energia will be able to sell these shares, or borrow against them, raising additional much-needed capital. If we did an IPO within a year or so, Energia, owner of 60% of MirCorp, might reap another $50 million.

Finally, Energia was a private company with its own shares of stock. Energia understood that if MirCorp did well, the shares of Energia should appreciate. This was in fact borne out just a few weeks after Goldin's public lamenting, when Oppenheimer & Co, the well known securities firm, informed Energia that they were now valuing the illiquid Energia stock at $50 a share, because of the profit potential of MirCorp. This increase in the stock price from $25 a share raised the valuation of the Russian space company roughly another $250 million. That's another $250 million that could be used as collateral for loans to pay for building rockets for International Space Station.

Energia's management understood the company could benefit in multiple ways from the deal. Through this prism, the MirCorp venture could be worth anywhere from $20 million to $300 million over the next year. That places the $65 million charged NASA in a respectful position. Did Goldin really not understand this?

Rohrabacher understood what Semenov was trying to accomplish. So too others in Congress and in NASA. So too people watching us from afar. Supporting emails, several thousand in the first week after the London signing ceremony, poured in from all across the world, including about a dozen from NASA employees, most of whom included their names. My favorite was this one:

At last!! That is all I can say and a big thank you to everyone involved on the project. For so long I have dreamed of commercial openings in space and always expected that it would arrive from the capitalist driven, democratic bound USA. Ironically, this huge stop for mankind has come from their counterparts in Russia with help from a modern day entrepreneur. This partnership between MirCorp and Russia is beyond doubt the best thing for the future of exploration of space. This will prove fresh challenges and new impetus to the space "race."

-France-

Section Four: Scheming Politicians, Star-Studded Customers 2000-2001

Editor's note: Why is it that a nation that is currently communist (China) and one which was communist for 70 years (Russia) seem to be much more able to commercialize certain on-orbit space activities than does the US - a country that has had a capitalist economy for several centuries?

-Keith Cowing, Internet site NASA Watch, September 11th, 2000

Chapter 27: Conspiracies and Boycotts

Noordwijk is a pretty coastal town with a wide main street that hosts high rise hotels and expensive restaurants with sweeping views of the North Sea. The wind is a constant and in the spring so to is the rain. A block in from the sea, behind the glittering wall of hotels, is the typical Dutch town. The clean small streets and pedestrian malls filled with bicyclists and pedestrians as numerous as the cars. Gert had arranged in town both an office and apartment. With some apprehension, it seemed the sort of place I could live for the foreseeable future.

The pleasantries of Noordwijk were at the moment lost on me, as I stood in the local branch of the ING bank staring in disbelief at the branch manager. I knew what he had said, as with all the Dutch his English was perfect. Nonetheless, I was forced to ask him to repeat. "We must decline having your company open an account with us," said the earnest young man, about 35, with short cropped blond hair, and a clean starched shirt with a hip shiny black tie. I must admit to taking a moment to understand. "You don't want our business," I finally stammered.

"That is right."

"And why?"

"I have seen the reports, its news on all the channels of our television. Your company is accused of taking rockets from NASA and disturbing the ISS space station project, which is important to us here. We don't want to be associated with this sort of problem." I asked to speak to his boss. An appointment was arranged for the next day, which I didn't keep. And then I was out in the wind, heading to the appointment with the landlord, with no local bank account.

The lack of a local bank account was no problem as the rental manager didn't want our business either. But at least he shed more light. As the manager for the major office building in Noordwijk, he dealt regularly with the aerospace contractors. That very morning the manager had received a call from the official at the European Space Agency who handled local business contracts. Bluntly, ESA had informed the rental manager that MirCorp was not welcome in the building.

No office. No bank account.

That forced us into plan B, which was to rent an apartment but keep Andrew in Montreal, and Chirinjeev in Chicago and Gert Weyers in Moscow and me flying between where needed. It was not a good solution; there were lots of dropped communications and misunderstandings. Kristin Oland in Washington became the MirCorp switchboard, the single point of contact for everyone. Many was the night while in Moscow or Europe or Asia I would receive yet another sarcastic email with a triple-edged message from Kristin-as she sardonically evaluated the unfolding situation. The aloofness of Walt, the eager enthusiasm of Chirinjeev, the lack of understanding from Andrew just how cutthroat was the business, and god only knows what she said about me to the others.

During that spring and partly into the summer, until we finally found an office in Amsterdam, I became a de facto expert on Internet cafes. At the time, personal computers were scarce outside of the United States and so too quality Internet connections. All over Europe entrepreneurs were commingling in a bar or a clothing store or whatever, with access to the Internet. That's how I did MirCorp business for months.

In Moscow I was living in the Savoy Hotel and doing most of my work at an Internet Café called "Baza 14," located on the second floor above a hip clothing boutique. There was a small bar and a crowd of young Russian students mixed in with some tourists. Especially late at night, when the business center in the Hotel was closed, there I would be, writing to investors and would-be partners and co-workers while impatiently waiting at the bar were students eager to talk to new friends on the local chat boards. It was not a good situation, but there was little choice. E-mails were impossible to send from Energia where I spent most of my days, as Semenov had never approved use of electronic mail by his workers. Even the fax machines were kept in a locked room down the hall from the General Director. So there I was at Baza 14, often well past midnight, drinking beer and writing and reading emails over the slow, slow servers of Moscow. Often a full ten seconds was required for a single message to even open. With a hundred or so emails waiting, I was forced to prioritize to catch any sleep. Friends grew angry when emails were not returned, but there was little I could do other than suggest we speak when I was in the States. Later, when I bought my first Russian mobile phone, the loud music of the café forced me to find other locales to conduct business and be on the Internet while we scrambled to find a more suitable existence.

In Amsterdam and in the small university town of Leiden, where we rented a flat, the Internet cafes were located in the coffee shops. That meant spending hours in windowless rooms heavy with marijuana smoke, as heavy metal music pounded incessantly. I took to signing my emails from Amsterdam with a heartfelt "peace and love," to the irritation of just about everybody.

Perhaps the most dangerous was a café in the Middle East, where the young man behind the counter felt compelled to warn that his government monitored the email traffic. Appreciated the heads-up.

In a small town outside of Tokyo was the most interesting Internet location. The hotel, which lacked Internet, referred me a few blocks away. Inside was a locker room, and the young clerk motioned for me to take off my clothes. I did so, praying that this was indeed some sort of Internet café. The inner door opened onto a communal steam bath, in which Buddhist monks were bathing while several colleagues in their robes came and went into the sauna. Off to the side were two old computers hooked to the Internet. While the monks quietly spoke and steam rose all about, there I sat wrapped in a white towel, contemplating the ironies of the universe while working on my email messages.

I took away from these experiences the realization that computers are hardly fragile. I asked a manager at my favorite Internet place in Amsterdam how long the computers lasted, given their constant use in a room filled with tobacco, hashish and pot smoke. He looked at me strangely; perhaps it was the first rational question he had been asked. Then a shrug, two years he figured. I can say that if I spent more time in that place, I would not have fared as well.

Conspiracy in Outer Space

Nothing was coming easy and there seemed some sort of universal conspiracy against us. In Moscow, Energia was restless as the sun was exploding in an unusually active heavenly burst of solar flares. These huge streams of energy, equal to tens of millions of atomic bombs, spewed forth through the solar system and were causing havoc with communication satellites and dragging down the Mir quicker than expected. We would need another cargo ship re-boost months earlier than expected if the flares continued. Imagine, cash flow for a start-up company impacted by solar flare cycles.

Closer to earth, there was no progress on the tether project. The State

Department had yet to take any action as the months dragged on. Though filed in December, there was still no approval. Now code-named "Firefly," Rick had put John Jacobson in charge. Every few weeks we all received another progress report from either Rick or John or the Washington lawyers, all depressingly the same. The equipment had been bought, the tether was being assembled in the States, but as of yet no action had been taken by the State Department. What was really frustrating was that the Pentagon folks had been understanding. Agencies like DOD's Defense Threat Reduction Agency (DTRA) had reviewed the Tether Applications' proposed technology transfer and determined that the technical data was not sensitive, and had no munitions application. DTRA even referred to it as "1958 technology." So we had to believe that it was the State Department holding back on approving the license.

Derechin had been skeptical from the start. Indeed, he had even laughed during one teleconference call with the Washington lawyers. This was very strange behavior for Derechin-it seemed that he knew State was not going to act. Perhaps from a source at NASA.

Nor did the Washington lawyers appreciate the force of the political opposition. A meeting with Goldin had been sought in early January to explain the tether to the Administrator. Once the story broke on MirCorp, his office canceled the meeting, and yet, far into February the lawyers sought to reschedule. Zero was the chance that Goldin would meet with anyone connected with MirCorp.

A number of unpleasant truths began to rain down. Derechin and I believed there would be no tether. At the board meeting in December it had been agreed that the deadline for sending the tether hardware would be April 15th. If not met, the MirCorp effort was not a marathon but a dash. It would take the Firefly team longer to accept the situation, but finally all of us understood that the idea of pushing the Mir into some sort of holding pattern was probably not to be.

No angle was left unturned by Goldin's NASA.

In mid-March I had dinner with a very senior ESA official. Mum was the word over the first bottle of wine. When the second was opened some information was revealed. "You would not believe how we are being pressured, directly, from the top on the American side to our top people." Yes, I would believe. I pushed for details, but the director let slip only one data point, as he believed I already knew that all contract discussions with Energia by the European Space Agency had been put on hold, in an effort to force Energia to withdraw its own support for MirCorp. Anderson had believed that the Europeans would welcome MirCorp, especially the fiercely independent Dutch. The reality was more mixed. Privately, European space officials sympathized. But fighting against the United States on behalf of a struggling entrepreneurial firm was a far different equation.

There was one incident where the Dutch were wonderfully supportive. By summer Gert had located a very classy office in the heart of Amsterdam. One day there was a knock on the door, and in walked a Dutch government official. He informed Gert that a reading of the United Nations Outer Space Treaty suggested that if a company had access to a space asset, such as a space station, then the nation of that company was liable, as was the launch country. He was right. After some debate in the Dutch parliament it was agreed that the Netherlands would accept their obligations under the treaty, and language was signed into Dutch law recognizing their national obligations because of MirCorp.

Conspiracy Continues on the Earth

In late spring we had the business plan ready to roll. So too the KPMG valuation. After extensive prodding by Kathuria, who pushed hard enough to anger Eddy, KPMG valued our lease, with plenty of caveats, at almost a billion dollars. Remaining was the roadmap from the major international consulting for locating potential customers. Armed with all three documents, we would fan out and seek both partners and investors.

First the consulting report was a month late, then two months. It was now my turn to push Chirinjeev, and finally the upper management in Chicago yielded and one late afternoon the report was faxed to our Amsterdam office. It was one page.

The report from the major international consulting company on potential aerospace, pharmaceutical, entertainment and Internet customers for the Mir space station suggested we contact two companies-Lockheed and Boeing. No contact names were provided. Just the general phone number for each multi-billion dollar aerospace company, the very two that felt most threatened by our existence.

Digging with the help of a reporter, I discovered that Boeing was a huge customer of the consulting company, and the International Space Station contractor took exceptional umbrage at their consulting firm working with MirCorp. That's not the end of the story though. A few weeks later the bill arrived for over two hundred thousand dollars. The invoice was longer than the report.

The strategy to bring us down was broad and effective. Without the tether we would have no choice but to find additional funding within the next couple of months. That meant tens of millions of dollars from investors or customers. We lacked a professional roadmap to potential customers, given the refusal of the consulting firm to work with us-nor did they even warn that no report would be forthcoming. For months Gert was scrambling to find an office, not an easy proposition in booming Holland. Energia was suffering from the boycott on the part of the Europeans, and given the lack of time, it would seem that our only hope for customers would be from outside the government space agencies. But the negative press would no doubt scare off many companies. The entire space industry watched and waited to see if we would stumble before we even began.

We somehow persevered. It was not easy; the experience was like running through mud. But the situation changed for the better when we announced a firm April launch date for a two-man crew to the space station. That was a tangible step everyone could appreciate.

Chapter 28: The Flyboys of MirCorp

I woke far too early on April 4th, impatient for the ride to Mission Control. If all went well this morning, a manned Soyuz would launch to the space station, continuing the life of the Mir. By dawn the crew at the Baikonur launch facility would have been just about through their preparations for Soyuz TM-30, also known as the MirCorp mission. It would be a dangerous mission, living onboard the aged space station with its degraded capabilities. The concerns were too numerous to mention. No one could predict what would be found when the hatch was opened. There was the threat that the pipes would have cracked. Or that the leak would have reached some sort of tipping point and was now beyond the ability of the crew to repair. Or toxic chemicals might be leaking.

The ground controllers at TsUP had begun returning the station to life. The main computer had rebooted with no detectable problems. From what the monitors were showing, the atmospheric pressure from the leak had continued to drop at the expected rate. The temperature inside the station was sometimes frigid. Unknown was the impact on the equipment, and only sending the crew would answer the question as to whether the Mir could still sustain a working crew. The leak was the biggest public relations concern. It had become the source of endless late night jokes, when it was reported that the crew had used duct tape to fix a host of space station prob-

Never has a manned space mission been so identified with a private company. From the manufacturing plant to the launch vehicle to the banner onboard the Mir space station the MirCorp logo was prominently displayed. Reactivating the dormant station was a difficult mission for (left) Alexandr Kaleri and Sergei Zalyotin. Both men believed in allowing tourists into space and dismissed the moniker of "space babysitters" thrown about by the majority of cosmonauts. Photo: MirCorp

lems, including the leak. As the date of the mission approached I went to Victor Legostaev and asked him to promise that when the leak was found, as I was sure it would be, that no one was to use the words "duct tape."

Say flexible plastic, I pleaded to the senior engineer. Say space-age fabric, but please, please, don't say duct tape.

For more than a year Semenov had argued that much could be learned by extending the life of the station to 15 years. This was the expected life of the International Space Station, and Semenov tied his dream of keeping the Mir aloft with the offer to share with NASA what was learned. Semenov made much that this mission would teach the whole world, including NASA, how to undertake mid-program repairs, whether it was the Mir or a mission on the way to Mars, or the ISS in a decade. This reasoning was rejected by NASA, which refused to become involved in the nuances of keeping the station alive while unmanned and then bringing it back to life. Among all the missed opportunities by NASA, this one may loom large when the ISS ages.

Joining me this morning was Chirinjeev Kathuria. We had arrived together from Amsterdam. Of course Walt was absent, but certainly not out of our thoughts. The need for further funds was dire. No one was paying Energia. Not NASA. Not Rosaviakosmos and not MirCorp. We had paid millions, but to continue building the vehicles for future missions more was needed. The next few Soyuz and Progress vehicles, for either Mir or the ISS, lay in different stages of development on the production lines. At a Paris meeting, Semenov and Legostaev had offered Anderson a percentage of the production line in return for an additional $10 million. It was a profoundly radical offer, but Anderson seemed distracted. He was not his usual boisterous self. Anderson asked several questions, but made no promises, instead saying he needed to think it over. So it was not just the future of the Mir that was in doubt, but also that of Russia's commitments to International Space Station. Funds from somewhere, from someone, were desperately required.

Waiting at Mission Control for our arrival was the ever present Alla Botvinko, our link to Energia. A determined woman in her fifties, Alla struck fear in many a Russian space official as she simply did not take no for an answer. Alla would barge into a meeting to get documents signed or push Semenov to provide more information. Men were just men, she would shrug. Alla informed me that all was on schedule while pulling from her bag the required papers for the security guards. Quickly we were inside TsUP, walking up the familiar marble steps beneath the stained glass mosaic of the three Russian space pioneers. At the top of the stairs was the chief of public relations, Vsevolod Latyshev, who had greeted me in December 1989, when I had arrived to witness the Payload Systems launch. We shared a good laugh at our mutual surprise that we were meeting again.

Waiting on the second floor balcony, overlooking the rows of computers and mission personnel, was dozens of reporters. The launch was scheduled for just after nine in the morning. Strictly following protocol, the State Commission had met and approved the final launch preparations

At about 4:00 in the morning fuel would have begun flowing into the Soyuz. There was the traditional final meeting with Semenov questioning whether the crew, and the backup crew, was ready. At about 5:40, in a ceremony practiced since the dawn of the Space Age, the two cosmonauts would stand at attention at the same spot where Gagarin had stood and report their readiness. Then I knew it was onto the crew bus, followed at a respectful distance by colleagues, family and reporters. The motorcade would travel from the Soyuz processing plant to the launch pad two miles away. The procession had undoubtedly stopped once, to allow the cosmonauts to get out and pee on the front tire. Why? Because Gagarin had asked his bus to stop so he could pee. When Norman Thagard began his training for living onboard Mir, Legostaev and all of the Russian community nervously wondered if the Americans would make fun of the tradition. They did-but Thagard semi-honored the ritual, standing by the tire but refusing to pee. The next emotional crisis for the tradition-conscious Russians came with Shannon Lucid. Would she also respect their customs? The tough Oklahoma astronaut did indeed, but moved to the far side of the bus, away from the prying eyes.

Minutes later the buses and cars would have continued on towards Area 1 where the Soyuz stood ready. Nothing is edgier than a rocket armed for launch. The tons of machinery strained and pushed like an impatient racehorse. For the cosmonauts it was the finish line for the months of training, studying, testing, memorizing, exercising, and self-politicking, and the start of a journey few can imagine.

The crew of Soyuz TM-30 carried with them the hopes of the embattled RSC Energia. When news came that the MirCorp funds had arrived, Ryumin immediately authorized four men, the primary and back-up crews, to assume first priority in Star City training. Rumor has it that Sergei Viktorovich Zalyotin and Alexandr Yuriyevich Kaleri raced into the astronaut training facility and pushed out the NASA ISS astronauts.

Zalyotin was the boyish looking 38 year old commander on his first mission. With a twinkle in his eyes, he was always ready for a joke or quick retort. Kaleri was quieter, a 44 year old Energia astronaut, a veteran of two previous flights to the Mir. Both had volunteered for this unusual mission, as there was concern among some cosmonauts that being associated with MirCorp might be detrimental to future flight opportunities. Not that Energia management did anything to hide that association. The training area and production facilities displayed huge MirCorp banners and our logo was clearly emblazoned on the side of the massive rocket.

There was optimism that the audacity of this mission had been understood even in the hallowed halls of the Kremlin. In early March Vladimir Putin had unexpectedly visited Star City. The President spoke with Zalyotin and Kaleri and promised support for Mir. "If there is a possibility to save it, this should be done," said Putin. "And I see such a possibility." Semenov and Legostaev were skeptical this was nothing more than the promise of yet another politician, but one never knew what a new political leader might do to show solidarity with his aerospace establishment. I saw Derechin a few days later and he was visibly surprised. There had been no warning by Putin and his entourage; they had simply showed up at Star City. Derechin worried that if something went wrong with our mission, now Putin would come down on Energia for embarrassing Russia. Nothing like a little more pressure.

The van with the cosmonauts should have reached Area 1 just over two hours earlier. After a few moments Zalyotin and Kaleri would have climbed out and walked over to the elevator, to take the ride up to the capsule. But first the two would turn and wave, and excitedly the crowd would wave back. The pre-launch scene for Soyuz is reminiscent of the early barnstorming flights of the aviation flyboys. Some in the crowd walk right up to the base of the vehicle. Kids run on the grassy field in the shadow of the fully fueled rocket. Dogs bark. Wives and girlfriends wave and cry. Space officials huddle, trading industry gossip. Once the cosmonauts reach the top of the rocket and are placed into the Soyuz capsule, the onlookers leisurely return to the cars and vans and pull back to the viewing area.

Here at Mission Control in the town of Korolev there is the usual steady and calm atmosphere. More media for this mission perhaps, certainly more foreign media. No matter the circumstances, Victor Blagov the deputy director of the Center and Vladimir Solovyov, the director and former cosmonaut, have been through the drill countless times before. I've known Blagov for years, he was there for us at the Mir Symposium and we remained in communication ever since. He's always been a steady force no matter the tension of a launch or space station crisis. This morning Blagov took a moment to come up to the balcony to brief us on the flight status. All seemed fine.

Derechin hit me with the latest news on the poor state of the Soyuz production line, and pushed Chirinjeev as to whether more funds will be coming soon. "Oh, sure," promised Chirinjeev, "I'm planning a liquidity event in June." Moments later I needed to explain to Derechin that a "liquidity event" meant he would be selling shares of stock to raise money for new investments, including another round in us. Why not sooner? Demanded Derechin. A good question, but my mind was focused on the upcoming liftoff.

Botvinko and Latyshev led us to the front of the balcony to wait out the remaining moments. We walked through the crowd of well wishers, past the family members of Russian space officials, contractors, experts from Energia and the other organizations. The actor Vladimir Steklov, the one scheduled to shoot a film on the Mir, was holding court, surrounded by Russian reporters. Steklov seemed a well-meaning guy who just shook his head at the difficulties of working with Hollywood types. Later I would have the chance to speak with John Daly, the British producer who had promised the outer space film would take place. He told a tale of agreements never honored, of stars with fickle attention spans and murky Russian intermediaries demanding money. If MirCorp was successful, he flatly stated, he would try again.

Chirinjeev certainly attracted attention everywhere in Moscow, including at TsUP, wearing his deep blue turban. Some, mostly women, would approach and ask about his Sikh beliefs. Not that the attention was always so welcome. Several weeks later he would be attacked by skinheads just outside the Savoy Hotel and saved only by the intervention of a waiting limo driver and a handy baseball bat.

When we arrived at our seats in the front row, I noticed that one of the television cameras was focused solely on us, discreetly placed on the opposite side of the balcony. "Listen to me," I whispered to Chirinjeev. "See that door over there?" He nodded. "In case the Soyuz explodes the media will be hoping to catch our reaction. Not from me." I grimly promise. Chirinjeev's wide eyes grew even wider. "If that happens, don't show any emotion. Just look at me and we'll slowly walk to the door." The door I knew led to a private viewing area. "You understand Chirinjeev?"

"I'm sure all will be fine, Jeff, right? Asked this young and wealthy man who had probably never suffered an unexpected setback.

"Sure. But you got it? Remember, no emotion, and follow me." I didn't want us trapped by reporters if the rocket exploded.

Now nothing to do but wait. With us were Botvinko and Derechin and Latyshev. The communication between the crew and the Baikonur control room crackled to life. The moments before launch are filled with eagerness, anticipation and nervousness. I think of the Challenger crew-fear has me drowning in negativity. What if the Soyuz does explode? Dan Goldin would blame MirCorp, no doubt. And it would be decades before a privately financed manned mission could again be attempted. Think positively. I look over at Chirinjeev. He is happily taking in the whole scene, the mission specialists, the reporters, the huge screen, with the green trajectory map where across the top is the countdown to launch time.

I hear a voice declare all is ready and then a pause and then, liftoff. I picture the Soyuz engines bursting into the blue-white flames as the rocket clears the tower. The sounds of exploding rocket engines differ. When the space shuttle lifts off, dark waves of sound crash into the observer bleachers. It is a deep primeval roar, as if some sleeping dragon has awakened. The Soyuz engines are a cataclysmic hiss, shearing of sounds punching you in the gut. Love it.

The green trajectory map with its blinking points allowed us to follow the flight path. Rising faster and faster, the first stage separated on schedule. Some scattered applause was quickly dissipated, like newcomers at the opera who applaud at the wrong moments. Unlike at NASA launches, no applause is welcome until the crew reaches the safety of orbit. Now second stage separation. OK, good, the tension eased somewhat from the knowledgeable observers in the room. Eight long minutes after liftoff the blinking dot on the huge screen indicated the Soyuz was now roughly horizontal to the earth. Orbital insertion was moments away. Telemetry now showed third stage separation. Zalyotin confirms. The room bursts into applause and the cameras swing onto Chirinjeev and me for our happy reaction. "Congrat's" I say to the investor, shaking his hand. Then I turn to shake Derechin's hand. It is a stronger than usual grip from the veteran of dozens of launches. I realize he too must have been slightly apprehensive. Behind me I hear a puzzled Chirinjeev. "What, what happened? Jeff, did the rocket launch?" Derechin and I just stare at one another. "Is everything ok. Did I miss it?"

An impromptu press conference was held right on the balcony, giving Semenov and I a chance to speak with reporters. I immediately committed a faux pas, congratulating the crew on the success of the mission. Latyshev stopped me. "We don't congratulate the crew until the docking." I made a mental note to add that to my long list of "Russianisms." Never shake hands through a door. Never toast a launch before it takes place, to cite just two examples. Oh, and try and sign a contract on a date with good numbers.

Semenov's comments were unusually informative. The General Director called the MirCorp mission a "landmark" in space exploration history, given that the mission had received no governmental funds. And as I was continually explaining to reporters and Washington officials, Semenov wanted the reporters to understand that Energia was also contributing its own funds to this mission. "We invested 400 million rubles (about $14 million) in the new station," bellowed Semenov in his usual expansive voice. "And the state still owes us 570 million rubles (about $19 million) for 1997-99. We also invested about 127 million rubles (approximately $4 million) in the construction of the Progress and Soyuz spacecraft." So from his point of view, Energia had spent millions on the governmental ISS program with no return from the federal budget.

An unexpectedly warm moment took place during the press conference when Latyshev pulled from his jacket pocket a worn blue envelope that contained a note written in poor English from a Chinese student, along with some American money. The student had sent the note to the Russian government requesting his $20 dollars be used "for saving the space station Mir." Latyshev held it up for all the reporters to witness, explaining the contents in both Russian and English. Everyone laughed, but I was struck by how seriously Semenov and Latyshev took the gesture from the unknown Chinese student. Semenov recalled the incident to whomever we met over the next couple of days, joking that if a Chinese student could send money, then so could the Russian government. The gesture resonated deeply for these men who had operated for so long in the darkness of the Soviet system. Men like Semenov were usually hated by the "public at large." It meant much that MirCorp had touched young people in a way that Semenov understood he never could. If people were sending money (and we had received plenty of supporting notes and emails in Amsterdam), then it meant we were on the right side of that strange unexplainable phenomenon known as public opinion. More than likely this was a very novel sensation for Yuri Semenov and other space officials.

The schedule called for Zalyotin and Kaleri to spend two days in the Soyuz catching up to the Mir. Much of the MirCorp staff remained in Moscow during this time, reviewing our strategy for finding new capital. Immediately the impact of having launched a crew into space could be felt. Lenorovitz was reporting far greater interest from mainstream reporters, and a wave of emails poured into our office. In Amsterdam Gert Weyers was carefully sifting through the avalanche, separating out the serious inquirers from those offering congratulations. Like panning for gold, the hope was that at least one of the emails would be from an interested partner or investor responding to the news of a private company reopening the space station.

On April 6th we returned to Mission Control for the docking. The balcony viewing area was far less crowded, mostly space engineers and officials. Everything had gone as planned. The station's leaking atmosphere had been refilled to the desired 650 mm, so the entry would be less dangerous. Chirinjeev and I watched as the green light was given by Mission Control to dock, and the final several hundred feet now separating the Soyuz from the station were slowly bridged. The docking between the space station and the Soyuz is usually done automatically, but with about 30 feet to go, the capsule was drifting a few degrees so Zalyotin grabbed control and manually eased the Soyuz capsule into the Mir port. Memories of the earlier crash of the Progress vehicle into the Mir remained very vivid in everyone's mind, and so there is a collective sigh of relief when docking was verified. Off to the side fellow engineers congratulate Vladimir Syromiatnikov for yet another uneventful use of the docking hardware.

Another critical milestone in the mission has taken place; the last great uncertainty was now the status of the station. Would it be habitable? We would have to wait at least an hour before the hatch would be opened; so Chirinjeev and I decided to grab some food and coffee at the small cafeteria down the marbled hall. Alla Botvinko hesitated; Energia never liked having foreigners mingle in casual settings with their people, but there was little she could do.

Syromiatnikov joined us, followed by the other senior engineers standing by in case of any problem in their area of expertise. Most of the men I had known since the days of Energia Ltd., so the conversation flowed freely. The men around the small table leaned closely into my face, explaining quietly that the situation was dire for MirCorp and for Russia. More than a dozen Soyuz's sat in different stages of assembly while NASA was promising precious money if-and only if-the Mir was brought down.

The number being bandied about is $60 million "under the table." Semenov had been refusing, but none of the men knew for how much longer. It was not just respect for MirCorp that stopped Semenov from acceding to NASA's demands. By this point Energia's and also Russia's mistrust of Dan Goldin and the United States was as deep as the Congress mistrust of Russian intentions. "We announce the Mir is coming down and what stops NASA from apologizing that Congress won't approve the sixty million?" Explained later Deep Throatskii.

Syromiatnikov wanted to discuss the tether. The self-imposed deadline was just a week away. Obviously the tether was not arriving anytime soon. It was a great disappointment to this veteran space engineer. Here had been a chance to contribute a second novel piece of hardware to the world. The first was his docking system, the APAS, which was now the standard for Russia, the United States and rumor had it, soon the Chinese. A piece of hardware that could reduce the number of cargo ships necessary for sending fuel to a space station would have been a satisfying second contribution.

Returning to the Control Room there was an unexpected surprise as Ryumin and Blagov had come over to stand with us. Also surprising, the huge screens suddenly displayed live images of the crew. Transfixed, we watched silently as the hatch to the Mir was opened and the men drifted inside. Zalyotin proclaimed a "thank you to MirCorp" for making the renewal of the Mir space station possible. Ryumin shot me a glance. The live images, the public thank you from the cosmonauts, all were signs of appreciation from Energia to MirCorp.

The first images and the tone of the cosmonaut's voices made clear that the station environment was as dangerous as we had feared. The temperature had swung from being far too cold and was now surprisingly hot. I believe it was close to 80 degrees. We could hear the Cosmonauts reporting that condensation was everywhere. Having no time to waste, the men got to work immediately, inputting start-up commands into the station computers, and the live transmission blinked off.

Now was the time for celebration. The station was once again open for business, against all odds. I must admit to feeling pretty good. No matter the fate of the business model, the reopening of the space station struck a small blow against the idea of the fragility of space hardware. Maybe we seriously underestimate the robustness of our machinery, whether station modules in the unforgiving environment of space or computers in the pot-haze of Amsterdam Internet Cafes.

We moved into a small room for the celebration. Joining us were several officials from the European Space Agency, a representative from NASA, a senior Putin official, military representatives and Energia management. Among the polite banter, and while digging into the piles of smoked salmon and fresh cold cuts, Semenov was brutally frank. "You never trust us," he barked to the NASA official. "MirCorp does, and we will work with them." What to say? The rift between Dan Goldin and Yuri Semenov was deep.

After the toasts led by Semenov for the success of MirCorp, which I'm sure the NASA official found unpleasant, Ryumin indicated he wanted to speak alone. We walked into a side room, followed by Victor Blagov. It was the family room, where years before I had spoken with the Mir crew. Once the door was closed, Ryumin began a small speech and Blagov translated the parts I couldn't understand. "Under the terms of the contract signed by you and our General Director the resources of the Mir space station now belong to your firm," solemnly began Ryumin. "Therefore, we wish to understand who will be stationed here at Mission Control to observe our activities."

My mind raced to match the respect being shown. "First," I began equally formally, "the investors of MirCorp wish to thank you for completing this difficult first step." The men nod in response. "Yet, I am puzzled, gentlemen. Under the terms of our agreement, all technical responsibility rests with the Russian side. All business with the foreigners. Therefore, it is your responsibility to assure satisfactory operations on behalf of MirCorp, which is your company as well." Blagov smiled and translated for Ryumin. Both men now smile. "No onlookers?" Again tests Blagov. "Only your conscience" I reply, and we all laugh.

"NASA and ESA never trust us," angrily echoed the burly ex-cosmonaut of the comments just made by his General Director.

"MirCorp is your venture also," I repeat for maybe the hundredth time to the Russians. Ryumin shifted his feet and turned his massive frame to directly face me. There was something else he wished to say. "Now that the resources of the station belong to MirCorp, as the general director of MirCorp you are now in charge of station operations. If there will be no permanent oversight, tell us now your orders, Mr. Manber."

Jesus. What to say? "Tell me your suggestions," was my clever response.

Ryumin spoke first. "Find the leak. Patch it up. That sends the message that the Mir can be continued." Blagov of course disagreed. "Let's get going on some scientific experiments. That will calm all those who believe the station is no longer productive." Time for compromise. Look for the leak, I order. Find it. Fix it. At the same time, announce commencement of scientific work.

We agreed and shook hands.

Ruse, of course. They knew damn well what had to be done, but these veterans of much of man's exploration into space had given me the honor that morning of treating me no different than their own senior management. "Check the box as written with the bosses and move on as originally planned." I also knew a bit about what was about to happen. Working with the Russians for more than a decade had taught me a few of their tricks. Yuri Semenov's birthday was fast approaching. I knew the crew would be hard at work to give the General Director a birthday gift, the sort of gift only a space crew could give to their leader.

And so it was. On Thursday, April 20th, it was announced that the Mir cosmonauts had, at long last, found and repaired a small leak with a sealing plug. It was also announced "the leak-related repair and monitoring are not the only activities performed by the crew today. They also worked with the Biostoikost materials science experiment."

Few Westerners, I'm sure, noted that the announcement of the repair of the leak took place on the birthday of Yuri Pavlovich, born on April 20, 1935, in the town of Toropets. For those in the Russian space community, it was a satisfying gift to the stubborn and fiercely proud General Director.

Chapter 29: Citizen Explorers

The Los Angeles sunlight was blinding, especially after the months of time spent in the dark gray of Moscow or the steady drizzle of Amsterdam. I used to laugh during Sea Launch discussions when an arriving Russian or Ukrainian delegation stepped off an arriving Aeroflot plane at the Los Angeles airport dressed in winter coats and hats and squinting in the unfamiliar sun. Now it was my turn.

Chirinjeev and I were lost. We were looking for an award ceremony in downtown Los Angeles honoring the manned space program, and Tom Hanks for his role in Apollo 13. Right now, all the blinding white modern buildings looked the same. We walked into an ongoing reception and the entire room stopped to stare at the Sikh in his dark red turban. We stared back at a hundred couples, with the men all wearing Yarmulkes. Wrong hall, unless Tom Hanks was enjoying his Bar Mitzvah today.

The announcement of Dennis Tito as MirCorp's first Citizen Explorer made real in America the pent-up demand for consumer participation in space exploration. His mission created the foundation for the modern era of space tourism, and also caused a great deal of soul searching within NASA as to their belligerent stance against commercialization. Photo: MirCorp

We haven't come all this way to pay homage to the actor. In attendance this afternoon will be a Los Angeles financial executive named Dennis Tito. For the past several months Tito has been meeting with Rick Tumlinson regarding becoming our first citizen explorer, or non-professional, to journey to space. Tito heard Rick speak on MirCorp and had followed our progress before finally contacting Tumlinson. Their discussions had been carried out over the telephone, over meetings and had finally reached a maturity that warranted our talking terms and conditions to the businessman.

Also at the award ceremony would be the film director James Cameron, who had earlier launched his own quiet discussions with Energia and Rosaviakosmos regarding shooting a film in space. The discussions had stalled. Rosaviakosmos was suspicious of working on a film project, and Energia had just been burned by its attention to the Russian actor and his Hollywood producers, and wanted no part of any Hollywood scheme.

So too Walt Anderson. From the start of the project Walt recoiled at the idea of the Mir as an entertainment platform. His opposition was a huge surprise, as I had mistakenly viewed Anderson as a capitalist pure and simple. Soon after signing the lease I went up to New York to visit my friend Mike Macmillan, who runs his own public relations firm for financial and Internet companies. Mike explained that companies spend tens of millions of dollars to develop name recognition. We tossed around the value of the Mir space station. A source of jokes since the fire and the accident. It's association with the problems of the Russian Federation. There was no doubt that the Mir had a negative image. "But its name recognition is incredibly high," mused Macmillan. "In every country in the world, almost every consumer knows about the Mir. It is cheaper to turn around a negative image then to create a new one, especially with an avalanche of new products or customers or programs."

Mike pulled a piece of paper from the printer. We outlined a series of necessary steps to capture the attention of the average consumer. The drama of a commercial manned mission, or the idea of Mir vs. ISS or a David vs. Goliath. None of that really worked. We didn't want to picture NASA as evil. "The lease of the Mir itself becomes the message," I ventured. "The formation of a peaceful company, which brings the Russian know-how in building space hardware together with the creativity of American entrepreneurs, that becomes the message in and of itself. We work with consumer companies to build an image as a post-Cold War product."

Mike finished the thought. "This is not about a space station; it's about creating a brand that is a cutting edge, twenty-first century consumer product!" Added Mike, "It's an American brand but it pretends not to be."

I ran the idea past Tumlinson who let out a yelp. "Oh man! That's so retro. It's exactly what I've been thinking."

Anderson hated it. "I'm not investing millions of dollars to create a new pair of jeans. MirCorp is for the scientific community," he scornfully lectured. "I want something serious here, not a gimmick. There is tons of scientific equipment stored on the Mir.

My view was different. "We could appeal to those $200 a pair jean companies, and the software firms and the Internet dot-coms and every other company with an edge."

"You're playing into NASA's hands," countered Anderson. "Let's meet them head-on. We are a space company with a space station. As a consumer company they can ignore us." Obeying Anderson's vision meant we located as close as we could to European aerospace companies in Holland, not in London or New York or Los Angeles. We sent Andrew Eddy around the world speaking with space agencies and with those national organizations without a space program. But to my surprise, we had a nibble. One ISS government space agency, working with NASA, expressed support for also signing with MirCorp-if we survived into next year. That was politically huge, but we had to keep it mum and the value of the contract was not large. Had we ample funding from investors, or the tether to boost the station into a parking orbit, that path may have borne wonderful fruit.

No, it was Hollywood that beckoned. And coincidence or not, Tito too hailed from Los Angeles. Anderson grumbled about how Hollywood was a waste of time, where people only love to hear their own thoughts over two hour white wine lunches, but gave his approval so off we went to sunny Los Angeles.

James Cameron had taken my arm and was using both his hands to shake it. It's almost a Hollywood kiss-but not quite. "It's really a privilege to meet you, what you guys are doing is huge, and necessary." I'm smitten.

Before the award ceremony we spent some time with the director. His desire was to film an IMAX type documentary from the Mir, including a spacewalk. Cameron's interest in frontiers and pushing the technical limits in filming had already been demonstrated, having shot underwater scenes for *Titanic* using a Russian deep-sea submarine. Coincidentally known as MIR, the special submarine allowed for filming at a depth of some 12,000 feet beneath the surface. For his proposed space project, Cameron had attracted interest from Japanese sponsors and was fully prepared to dedicate the months at Star City necessary for his own training. The director envisioned a 3D movie, and it would later emerge that the Fox network was sponsoring the deal. If implemented, Cameron would become the first citizen explorer to undertake a space walk.

Cameron impressed me with his vision regarding using space as an entertainment medium. He spoke eloquently about igniting interest again in space exploration, something he was kind enough to say we at MirCorp were also doing. Chirinjeev considered out loud that we might be able to find project financing for a Cameron project, putting us in a position to not just be a space hardware venture but truly an interface for creative projects. We promised to stay in communication. As we were leaving he asked us to keep our discussions quiet. Reluctantly, we agreed.

Before the ceremony got going the hosts also introduced me to Hanks, who was then in production for his upcoming film "*Cast Away*." Unfortunately, the actor was in character, his hair long and scrawny, with a full beard and with little to say. "He really, really has an interest in what you are doing in space," explained what seemed to be an eager assistant, "but he can't come out of role, you of course understand."

We fared better sitting next to Dennis Tito and his date. Like Anderson had done months before, the businessman blasted off rapid-fire questions about working with the Russians and their commitment to honoring an agreement. Tito was emotionally intense, a former Marine who founded a financial tracking service for pension funds known as Wilshire Associates, and later created the Wilshire 10000 Index, which tracked a large percentage of publicly traded companies. Using complex algorithms his firm was able to help understand macro financial movements.

Even when speaking in conversation, Tito's eyes blazed. This man was very tightly wound. He made clear that he had long thought of space travel and had approached the Russians a few years before, but the cost of a divorce had nixed his dream at that time. During the Hollywood ceremony he startled our table companions by introducing us as his enablers for his upcoming space mission. Great news indeed. The attendees were understandably shocked. Truth be told, it was a shocker to me. That MirCorp's first customer would be a space tourist, and not a research agency was pleasant vindication. It's like saying something over and over, despite the strong disbelief of critics, and yet it turns out to be correct. So many observers, in newspapers, on web sites, at industry gatherings, in the halls of NASA and ESA, had ridiculed the very concept of everyday people wanting to journey to the Mir.

Rarely have I experienced envy. Not flying in private planes, not sitting on rooftop gardens in mid-Manhattan or aboard huge yachts in the Mediterranean. But poolside at Tito's home, with its expansive views high above Los Angeles with the ocean sparkling beyond, leaves one with the question of just why the owner would leave this oasis for six months of military type training in Russia, and a touch of envy. That Tito was serious was without a doubt. After giving Chirinjeev, Rick and I a quick tour of his lovely home, he immediately continued the firm negotiating stance expressed to Rick. In 1991 the Russians had offered him a price tag of $12 million; Tito insisted he would not pay more this time. That was an issue solved with some unusual ingenuity. Over time, speaking with Derechin and with Legostaev, we hammered out the outline of a media package and a launch package, each with different price tags. The total package was worth $20 million. MirCorp would take the first $8 million in media revenues from his mission, so we could honestly say the deal was worth $20 million.

Important to us was the role of a space traveler like Tito. Rick had coined the term "citizen explorer," in the belief that this was not the time for space tourism. The word "tourism" implies relaxation and enjoyment, rather than risking one's life as a space pioneer. I wanted to make sure Tito understood the difference. The Mir was cramped, the cosmonauts would be busy, no guarantee for safety. "No room service?" Tito laughingly concluded, showing me full well he understood. Nonetheless we spoke for some time that afternoon and later over the phone, about how the Russians would view a rich American taking up the third seat. NASA was not that much of a worry. Even Dan Goldin would understand that Tito's funds would pay for the construction of the next Soyuz and frankly, the one after that, since it was partially built. That second Soyuz would be used for ISS. Absent any further funds from either government, Semenov would partially fulfill his governments' ISS obligations for the next year via the payment from Dennis Tito.

Tito was more worried about who else we were speaking to. He had seen us talking with James Cameron and there was indeed one other person very, very serious in flying. "I want to be the first," Tito flatly declared. "And Rick promised me." It seemed that Tumlinson had given assurances that Tito would be first in line. But, I countered; he had to be first to pay. This concern was the only weakness I could detect in the negotiations. Tito wanted very badly to be first.

The afternoon at Tito's home was invigorating. MirCorp's plans seemed doable and the huge problems receded as we spoke seriously of sending Tito to the space station. Chirinjeev was very good at addressing Tito's emotional concerns and, with his medical degree, could speak to the physical challenges of space flight. Not that much encouragement was needed. Tito had read Jerry Linenger's *Off the Planet: Surviving Five Perilous Months aboard the Space Station Mir.* The title sums up well that astronaut's view of his time aboard the space station and the strained relations with the Russians.

True to his Wilshire Associates roots, the ex-engineer had dug deeper. He studied the Soyuz, studied the launch rates, and understood the situation with the production line. Like Anderson, he too was bottom fishing, knowing that pouncing in at this moment gave him the greatest leverage. To my bemusement, Tito was also eager to show that he could behave just like Walt Anderson. He had read the *Wall St. Journal* article on Anderson's cutting through the mistrust, and wiring the Russians the seven million dollars. "I'll do the same," the financier promised as we sat by his pool. "Once we agree I'll wire what you tell me. Will a million dollars do it? I understand just how badly NASA has behaved with the Russians. I know we mock these guys when we really should be learning from them. Don't worry about me. I'll be a trooper. I'll salute and obey."

I left Los Angeles flying high.

Picking Candidates

The experience in selecting the first citizen explorer marked the start of the most schizophrenic period of MirCorp's existence. During the more than two months of the MirCorp manned mission our business model began creaking to life. It was a joy to behold. Despite the boycott, despite the shortage of funding, the business model seemed on track. Industry analysts and hesitant entrepreneurs had long debated whether there truly were wealthy individuals wanting to journey into space. The answer was an emphatic yes. Like Rene Anselmo launching his PanAmSat satellite before receiving regulatory permission, there seemed no better reality check than simply opening the doors for business.

Anderson's vision of being a low-cost alternative for space research also seemed to have merit. One ISS partner was desirous to work with us, citing the cost savings, the useful equipment already onboard the Mir and for sending a message regarding commercialization and NASA's choking of true competition.

We also experienced serious interest from the emerging Internet community on using Mir as a physical portal for space enthusiasts. A Hollywood icon like James Cameron wanted to film in space. A Los Angeles production house submitted a proposal to conduct a game show where the winner would fly to the Mir. Each of these deals would bring tens of millions in revenue and offer wonderful branding for MirCorp. Each would create unprecedented excitement towards space travel. Each should have been made possible by NASA and the United States space industry. None were. Instead each project when announced was mocked by the aerospace industry as a sign of a desperate Russian space program, while public enthusiasm ramped up for the first time since the onset of the space shuttle program way back in Ronald Reagan's time. But what was schizophrenic was that we were running out of time and money just while our audacious market assumptions were being proven correct.

If the solar flare cycles were reaching a harmful eleven year peak and dragging down the space station, these same bothersome flares may well have also been dragging down the stock market. Unbeknownst to us, the NASDAQ market had peaked on March 10th, 2000, and had begun the slide that would eventually be known as the "dot com" crash. This was just a couple of weeks before our launch and may be one explanation for why Anderson had elected to cancel his latest journey into Russia. It was Andrew Eddy who first voiced concerns. My attention was so focused on the day to day challenges I had overlooked the market ramifications. In the period of the week after the 10th of March, the market had fallen about 10%. From my perspective, Anderson was rich, right? His friends were rich, right? So what's a 10% decline, right? Wrong. Anderson was over leveraged, with his own stock holdings backing other Internet ventures which were collateral for other ventures. And Walt was no different than other paper multi-millionaires. Including Chirinjeev. Including their friends. At a recent meeting in Moscow, Anderson had spent much of the time staring at a scrawled page listing dozens of his securities. The "liquidity event" promised by Chirinjeev slipped from spring to summer and eventually to the fall and never took place. The value of the high-flying Internet ventures crashed. A $40 million dollar holding in one single Internet company tumbled to almost worthless, Chirinjeev later admitted.

We were also taking media hits that hurt our chances to bring in new investors. For every good solid story there was a zinger. In the spring NOVA, the PBS documentary series, produced a fantastic piece looking at both the International Space Station and the efforts of MirCorp to compete against NASA. It had Congressman Sensenbrenner suggesting we should use the Russians as contractors, since they honored their commercial commitments, but not their government-to-government obligations. Just as we had been screaming for the past five years. The NOVA documentary also included an interview with Walt who spoke of using the Mir for a variety of applications, including "silly advertisements." Despite the put-down, we received interest from major ad agencies associated with those "silly" advertisements.

Then there were the negative hits. In July we were blindsided by *The New York Times Magazine*. Freelance writer Elizabeth Weil had followed Walt, Chirinjeev and myself around Amsterdam, and then accompanied Chirinjeev and me to our launch. Whoever said all publicity was good never had an article like this on their company. I had been disappointed when Elizabeth pulled me aside at Mission Control and confided that the article would focus on the investors, and would not include me. I mean, after all, she was writing for *The New York Times*. But when the article appeared, I thanked my lucky stars.

American Megamillionaire Gets Russki Space Heap! opened with this description of Walt:

"Walter Anderson - a man with white hair and pale skin, square, gold-rimmed glasses and a physical presence so profoundly unprepossessing it's almost impossible to remember what he looks like. Anderson is 46 and worth almost a billion dollars. He lives in Washington - the city he grew up in, a city he hates; his hatred of the government is, as he puts it, ''personal'' - in an apartment adorned with a painting he commissioned based on a Smashing Pumpkins lyric, ''I am still just a rat in a cage.'' ''That's what we are,'' Anderson explains, ''rats in a cage. And we're going to gnaw through the bars because we've got about a 30-year window here, and we'll starve if we don't get out.'' The cage Anderson refers to is the planet earth itself, and he has taken it upon himself to ensure we get off."

The article only got worse. Depictions of Chirinjeev as a love struck amateur playboy. "Kathuria, however, the consummate sidekick, desperately wants to discuss his problems with girls. The situation has grown kind of dire in that he's already too old to advertise for an arranged marriage in the Indian newspapers: Sikh tycoon with lease on Mir seeks to marry same?"

The article was accompanied by campy cartoons poking fun of both men. Walt had hid nothing from the reporter, including our late night jaunts to the Amsterdam cafes where Walt would espouse his dreams for space travel over a bowl of hashish.

I didn't read the article until much later. Once Gert described it to me there was little motivation. It seemed an unprofessional attack, unless Weil told the guys from the start that it was a personal piece and everything was on the record. It looked like the two entrepreneurs had made the same mistake as Tumlinson, unable to separate personal from on-the-record conversations. Now I understood why corporate executives needed media handlers.

Like in a political campaign taking shots from all sides, we had all grown pretty thick-skinned, but the *Times* ambush hit pretty hard. The mocking of the two investors by the influential publication more than likely doomed any hope of finding new investors. Who would want to be associated with a venture so ridiculed by the *Times*? Tumlinson was suspicious; believing that there had been a deal struck with NASA, given that Dan Goldin had allowed himself to be interviewed. His participation on the record made no sense unless he was promised the piece would skew the MirCorp investors. Or maybe the other way around. NASA would provide total access to the *Times* if they would damage the investors. Who knows?

During the spring, while Zalyotin and Kaleri spent weeks cleaning up the Mir and conducting top-notch research for Russian organizations and one pharmaceutical company, Walt remained outwardly calm. One or two customers he reasoned, would provide us with revenue projections in the tens of millions, and certainly we could borrow against that. I was growing less sure, and wanted, frankly, to maximize revenue from the three candidates who were vying to be our first Citizen Explorer. There was Tito, a perfect symbol of MirCorp. A successful American business executive, a former NASA engineer and Marine. No one could question his reputation or his personal desire to fly. His age was a negative, at 59. And the medical tests had caught a problem, but it was solvable. Tito briefly resisted, telling the Russian doctors that he had just completed a full checkup at UCLA and nothing had been detected. "Did you tell the doctors you would be spending a week in zero-gravity conditions," astutely asked Derechin? Tito thought for a moment before conceding that could make a difference. A routine operation at UCLA made him flight ready. He interrupted a meeting I was having at Energia, shouting into the phone from his hospital bed to tell Derechin that he was now "fit as a horse," or words to that effect.

One unfortunate negative was that we didn't believe Tito would end up as an investor in MirCorp. Despite the press accounts that always err on the wealthy side, he was throwing a good chunk of his wealth into the space ride. He might introduce us to his friends, as on one occasion he put me on the speaker phone with Michael Milken, the former junk bond king. But even early on it was clear his desire for space travel was personal. Flying into space was not an investment; it was a journey of personal self-exploration.

The second candidate was also wealthy, and wanted to remain anonymous unless selected. He insisted on being first, but again, there was little chance of moving him beyond being a passenger.

And then there was James Cameron. The director of the mega hit *Titanic* had a genuine long-term interest in opening the space frontier to moviegoers. He was enthralled with using IMAX and 3D techniques in ways never attempted. Cameron offered a stirring story, a famous Hollywood celebrity who would ride up to the Mir and bring alive the drama of living in space. Cameron had secretly come into Star City and undergone medical testing. Given a clean bill of health was an important first step for his sponsors, and the filmmaker liked what he saw at Star City.

For me, the choice was obvious, just like it had been far earlier when John Denver had made his offer to NPO Energia. The problem was that the funds were not locked in for Cameron. His sponsors were committed, then not committed, then perhaps in a year. Cameron also pushed to ride first up to the Mir. Why not? Do the film and then allow individuals to fly. So just announce Cameron and hope that locks in the financial package, was that the right way to go?

There were others that wanted to fly. An Italian named Carlo Vibert was of particular interest because he had already trained with the European Space Agency. The Italian space engineer was assembling a team of media companies from Turin to sponsor his mission, much like the bike sponsorships so common in Europe. His plan never progressed, but is worth mentioning because of the response from the European Space Agency. I was called the day the story came out by one well-known ESA official who threatened that they would charge MirCorp and Vibert for the cost of the earlier Mir training supported by ESA. Then, a few days later, there was a call from an ESA official I didn't know warning that Vibert had been let go in "bad circumstances" and this would all come out.

During May and early June we went back and forth internally debating the value of the three serious candidates. As a team we decided that an announcement would have the most impact if made when the Soyuz TM-30 crew returned. Partly because the world media would be focused on us at this time, and also to deflect the growing realization that Walt and Chirinjeev were not contributing more funds for a future mission.

Cameron was not waiting for our decision alone. He had his own ties into the Rosaviakosmos, which he used to promote himself. Not that this helped either his cause or ours. Foreigners had a hard time understanding that the management of Rosaviakosmos did not promote, nor wish to promote, a mission like Cameron's. We would see that later with Mark Burnett, the reality TV producer, and with Lance Bass from the boy band 'N Sync.

The concerns on the part of Rosaviakosmos were not solely based on turf protection or a desire to please Dan Goldin. There was also the deep-rooted fear that a rich private citizen could well turn out to be a nut or go crazy during a mission. Teams of psychiatrists and psychologists studied potential astronauts and cosmonauts for years before deciding who was mentally fit for space travel. The Japanese reporter Toyohiro Akiyama had chain smoked during his training, which made the Russian specialists very wary of his behavior in space. What if a business executive freaked halfway through a mission? I figured we needed a series of non-professionals to adequately fly into space before the professionals would relax.

Given Rick's promise, I eventually threw in the towel and reluctantly agreed that Tito should be offered the first slot. But I insisted on extracting some additional fee for the privilege. Or at the very least, some concessions on his relationship with MirCorp, his speaking for us after his mission, and so on. Walt was adamant against this sort of negotiating. "This is not how I do business," was his reprimand. Later, he would rue not setting some basic conditions.

Time was again short, as usual. The crew was scheduled to end their mission on June 16th. This was an extension from the planned 45 day mission. The extra time was a consequence of how smooth things were going onboard the space station and also to try and give us more time to raise further funds. But even the mid-June landing gave us only a few weeks to prepare a contract and open an escrow account with Tito's funds. I called Dennis and informed him that we would be honored if he would take the next third seat. He was ecstatic, to say the least. The contract was for $20 million, with $12 million in cash, and $8 million in media rights, as has been reported in some media outlets.

But, I explained there wasn't enough time to have the contract signed before the press conference for the returning crew. That we could announce something later. Tito pressed, wanting to lock us in to a public announcement. He could hear the hesitation in my voice. Having been burned so many times I sure didn't want to make a fool of myself in front of the press and announce Tito only to have him change his mind. "I swear on the lives of my children I will come to Moscow, sign the agreement, and wire the deposit."

Taking a big sigh, I invited him to Moscow.

Branding Tito

The introduction of Tito had to be perfect and not a repetition of MirCorp's initial media stories. The odds were already moving against us, but good news coverage might still attract investors. The focus had to be not on our original investors but on Dennis Tito. Lenorovitz and I huddled to figure out the best way to present the Citizen Explorer. We decided that Tito as a symbol of MirCorp was the right way to market his trip. Like MirCorp, Tito was a bridge between the two once-competing nations. Like MirCorp, Tito showed that the space race had come to a close, and it was now time to open space to everyday travelers, including paying customers. We decided that in press accounts and background briefings Lenorovitz would frequently use the description of Tito as an ambassador. We ran it by Tito and he approved.

Then there was something else we wanted to accomplish. I'm not sure whose idea it was, but we elected to push Dennis Tito as history's first "Citizen Explorer." We of course knew that two other private citizens had journeyed to the Mir via Energia. Toyohiro Akiyama was a crewmember of Soyuz TM-11's December 1990 mission, a reporter whose mission was paid for by the Tokyo Broadcasting System. Then a British woman named Helen Sharman spent a week on the Mir in 1991. The 27 year old Chemist obtained her ride to space from a national competition in Britain that attracted more than 13,000 applicants. Both were true space pioneers. But we wanted to do something more than announce Dennis Tito as the third commercial space explorer. We decided to use Rick's "Citizen Explorer" to our marketing advantage, and referred to Dennis Tito as the world's first "Citizen Explorer." Energia was uncomfortable, feeling it diminished their earlier accomplishments. As such it mirrored the controversy when NASA unsuccessfully sought to name the new international space station Alpha, as if it was the first space station.

It took Legostaev personally to approve of our marketing approach. Confronted yet again with the Western quirks for branding and marketing, Professor Legostaev reluctantly agreed. Honestly, we were surprised how well it caught on with an excited media. Few reporters bothered to do their own fact checking and report that Tito was the most recent private citizen to journey to space. Later, many reporters would drop the "citizen explorer" moniker and use only the term "space tourist" to describe Tito. But his description as being first remained, and Tito became imprinted in the public's mind as the "first space tourist." Later, when tourists flew in flights brokered by Space Adventures, the reign of the space tourist was described as beginning with Tito, never before.

When pressed by some knowledgeable reporters, and remember this was before Wikipedia and search engines, we explained without hesitation, as did Tito, that he was the first to pay for his own ticket. True enough. But historians now note that Edmund Hillary and Tenzing Norgay together were the first to climb Mt. Everest, correctly drawing no distinction between the explorer who paid, and the explorer who was paid. Nor does history put an asterisk next to any explorer who had corporate sponsors rather than paying their own way. The mode of payment as a consideration for media attention went unnoticed in the space industry until we promoted Tito.

Helen Sharman and Toyohiro Akiyama have faded from the story of space tourism, in part because they were non-American explorers, and U.S. reporters remain myopic against most things foreign. And in part because of the huge publicity the Tito announcement garnered. My apologies to both Mr. Akiyama and Dr. Sharman for the role I played in minimizing their historic roles as the first man, and first woman, non-professional space travelers.

Tito is Announced

Sometime early in the evening Moscow time on June 15th the cosmonauts Zalyotin and Kaleri floated over to the Soyuz capsule that waited patiently to return the men to the earth. Certainly there must have been thoughts as to whether this was the final expedition in the fifteen year life of space station Mir. The men collected precious research cargo, gathered up the MirCorp promotional material signed during the two months in space, and made their way into the Soyuz. A video camera recorded the hatch closing for the last time. The crewmembers waved to the camera, the hatch closed, and a MirCorp logo pasted onto the forever-shut hatch swung into view. Our logo was the final image seen from the space station.

Hours later on the 16th, the Soyuz streaked safely down to the earth, delivering the men to a perfect landing in Kazakhstan, after a mission of 74 days. I'm not sure what Anderson was thinking when our mission ended. Whether he savored the moment or was caught up in the collapse of his financial net worth.

Two days later the traditional press conference for a returning crew was held at Star City. The crew was present, of course. So too Vasily Tsibliyev, now deputy head of the Gagarin cosmonaut training, and one of the crew members onboard Mir when the cargo ship collided, almost killing them. The room was crowded with reporters. After the crew finished their post-flight briefing, Derechin, Tito and I took our positions at the table. It was a sweet moment to announce that Dennis Tito had been chosen as MirCorp's first citizen explorer, and would fly to the Mir during the next mission. The reporters exploded with questions. While I introduced Tito, I had the feeling that not a single reporter in the room was looking at me. All eyes were on this wealthy American businessman willing to pay millions to risk his life on a voyage to Mir.

The questions to Tito were difficult and at times rude. One reporter asked whether he had read Jerry Linenger's book depicting his Mir experience. Others asked whether he trusted the Russians. Whether he would take life insurance (he would consider it but his children were fine regardless). The price of his flight. Was he really comfortable hurting NASA's space plans? And, above all else, why? That was a question we all sought to answer. As usual, Derechin was hounding me to better understand the customer, in this case Tito. Why this successful financier, soon to turn 60 years of age, was risking it all for a space flight. During his cosmonaut training there came a moment appropriate for me to ask Tito the question directly. He usually answered in vague expressions of fulfilling his life's dream. This time he looked me straight in the eye. "When you're my age, and want to attract good looking Los Angeles women, look at what you hafta do." This was a tough guy to understand.

For everyone the announcement of Dennis Tito seemed to answer the question of whether MirCorp would continue. In that sense it was a huge success. This was a high stakes poker game being played by our investors, by Energia and by Tito. Each was angling to achieve what might seem to be the same objective, a commercial Soyuz flight with Tito onboard. But the devil was in the details.

I carry from that press conference a special memory, and it had nothing to do with Dennis Tito. A reporter asked, as the final question to the Soyuz TM-30 crew, whether a huge solar flare that exploded out from the sun on their last day in orbit had been observed. It was a timely question, since the high number of solar flares continued to cause both the NASA space station and the Mir to be dragged down unexpectedly. A cargo Progress ship loaded with extra fuel had been sent to the Mir station in April, which cost us precious funds. Commander Sergei Zalyotin thought for a long moment and then quietly replied "words have yet to be invented in our language to describe the beauty and majesty of what we saw in the heavens that day."

I thought that a wonderful end to the final press conference for the final crew of the Mir space station.

Chapter 30: Reality TV Space Race, Our Future Rests with Game Shows

Rarely does reputation live up to the reality. Not so with the Hungry Duck, the legendary bar in the center of Moscow. On any night the place was packed with university co-eds wanting to let off a little energy and by midnight women could be found dancing on the bar, dancing on the tables, dancing in the cramped litter-infested hallway. The Duck was in an apartment building just a few short blocks from the Savoy Hotel. It was down one of those Moscow side streets, crammed with kiosks and kids hanging out in all weather drinking beer. Tonight we were showing the Duck to someone who certainly appreciated the value of a high energy show. Mark Burnett, the producer of the CBS *Survivor*, was our guest. Also making the trek from Los Angeles was Conrad Riggs, his business manager. Walt and Chirinjeev had also flown to Moscow.

There was no one riding higher than Burnett in the entertainment industry in the fall of 2000. And here he was, taking the time to come to Moscow and meet with Energia and the other Russian space organizations. As usual, it was an extraordinary clash of cultures. For Burnett, he saw the value of turning Mir into a positive icon of a new world where Russia and America were no longer enemies. For Russians, there was the usual suspicion towards a

Hollywood and celebrities saw enormous potential in space-themed projects. NBC and *Survivor* producer Mark Burnett were the first to ink a deal. Lance Bass of boy band 'N Sync, seen here with MirCorp's Gert Weyers, failed to attract the sponsorship necessary to pay for his Soyuz ride. But interest continues to this day and it is a market waiting to be tapped. Photo: MirCorp

market that usually insisted on payments based on the revenue. Burnett was proposing a reality show called *Destination Mir*. The series would follow a group of would-be cosmonauts from space training at Star City to launch pad over 13 to 15 episodes, culminating with a live broadcast in which a winner would be picked and sent into space. It would be an unprecedented introduction of the Russian space effort for most Americans, and probably a great show to boot.

After a few days visiting the Russian space facilities, Burnett had in mind the opening shot. He was telling us his vision while out of the corner of our eyes we could watch the first two Russian co-eds jump onto the bar. Burnett remained focused intently on the storyline. "It's quiet." The producer shouts over the noise of the Duck. "We are eye level with the tops of the birch tree forest of Star City. Then, suddenly two huge Soviet military helicopters break over the tree line. It's loud, scary even." The producer stops for maximum effect, or to let the club DJ stop his shouting. "The choppers swoop past us with blazing speed and come down hard in a clearing. Out pour the contestants. The battle to be chosen as an astronaut begins in the stark birch tree forest of Russia."

Mark Burnett is a tall ruggedly handsome Brit who had followed an unlikely path to his fame and fortune. Born in London, he served in the military, including a paratrooper stint during the Falklands War. Then came a decade of odd jobs in Los Angeles. When we met, he was still shaking his head over the fortunate turn of events and unsure of his future path. He explained how he had personally rescued the *Survivor* program and reaped the resulting rewards. As Burnett explained, he had alone met with advertisers in New York, persuading enough to come on board and support the then unheard of concept of reality television. In return CBS willingly gave him a large slice of the advertising revenue. No one really imagined the show would be a cultural phenomenon. But it was, and rumors were that Burnett was pulling down ten million dollars in the first season from his share of the advertising revenue.

Both he and Conrad weren't sure how long the *Survivor* series would last. "We figure two, maybe three seasons at best," predicted Conrad. Every Burnett needs a Conrad Riggs. Riggs was sharp, always mentally jabbing, seeking the optimal advantage, leaving Burnett to float high above the business. "I'm forty years old," Burnett mused, "and the question is whether I really want to be doing this for the rest of my life. "Look, I could propose a reality show about a shoe, and right now I could sell it to the networks. But for the sake of my family and my reputation I want to do something more. And *Destination Mir* is that chance for me. I see this show as a way to highlight the drama and excitement of space travel in a package every American can understand." As usual, the Russians were not convinced. So the hottest television producer in LA, if not the world, had to journey to Moscow with a tape under his arm peddling his work. Not that the Russians were the only ones unaware of Mark Burnett.

The dot-com crash was just one of the events that I had overlooked. For the entire past year little time had been spent in the States, usually quick trips to Los Angeles or New York, with a stopover in Washington. Kicking back and watching television was just not something I was doing lately. So when Chirinjeev informed me that he had made a connection to Mark Burnett, the producer of the CBS reality show *Survivor*, it made little impression. I had already opened discussions on a game show where the winner would go to space with another production company, and had held several meetings with the producers. We were probably just a few weeks from an agreement, at which time they would pitch the idea to one of the networks. There was annoyance at Chirinjeev's interruption. Tumlinson had pretty well dictated the terms of the Tito deal, I thought to our disadvantage. Energia had dictated the terms of Cameron and now, as we were closing in on a game show deal, here was Chirinjeev interrupting events with a new producer with the same idea.

A few phone calls quickly disabused me of the notion that the deal I was about to close would be equal to any Burnett production. From Los Angeles to New York the response was the same. If MirCorp could ink a deal with Mark Burnett, sign!

But the logistics of *Destination Mir* were difficult to imagine. How to allow tens of thousands of people to apply. Age limitations would be necessary. So too some medical parameters to screen those too ill, or too tall or overweight. How to further screen the thousands who passed the first review. And when to allow examination by the highly skilled Russian space psychologists would also be critical. So too the rules and traditions of space flight. Any member of a crew needed to be vetted beforehand, and it was always insisted that the entire crew should spend time together, train together and bond together. How to have that happen without losing the element of surprise was a question critical from an entertainment point of view.

Scheduling was another issue. Neither the Russians nor MirCorp would ever agree to financial penalties if a launch were postponed, as Riggs first insisted. Nor could we agree to a news blackout, giving the sponsoring network sole news rights to the launch. "Mr. Burnett is buying a seat on the rocket, not the entire rocket, the three seats and the launch pad," was Derechin's response whenever Riggs sought too much.

Earlier in the summer I had spent the three most excoriating days of my life locked in a windowless room with Conrad Riggs, and at times Mark Burnett and Chirinjeev in Los Angeles. Negotiated was every single possibility. Merchandising rights occupied two weeks of discussion and then one tough day of line by line negotiating. The back-end revenue, sharing of advertising revenues and payment schedule took up a day and a half of detail. And more than half a day on the role of MirCorp in the show, how to be listed in the credits, how many mentions per episode, whether our logo patch would be shown, whether any mention of MirCorp would be made by the contestants, occupied the final four or five hours of the hellish negotiations. I learned more in those three days on the business of marketing a space asset to the mainstream media than a hundred earlier discussions and proposals. Conrad Riggs was tough and nothing was left to chance. We had the assistance of a connected entertainment law firm but still the details were time consuming and numbing. But when the deal was done, we had a winner. Burnett then shopped the offer to each network.

In early August we received the sort of email that one does not easily forget.

Jeff,

I pitched CBS today as my fifth netywork (sic) after in order.

NBC, FOX, ABC, USA, and then CBS.

They all want it. Conrad will make a deal.

M.

By the end of August, the deal was in place.

The agreement with NBC was incredible, and would have resulted in huge revenues for Mark and Conrad, and extraordinary exposure for MirCorp, as well as a solid share of any merchandising revenue. Had the show run for 3 years, for example, we could have enjoyed over $120 million in revenue, according to the estimates from our entertainment specialists. The NBC announcement anticipated building relationships between the Internet, live television and space exploration, just as our business plan called for. Edmond Sanctis, the president of the NBC Internet venture known as NBCi gushed how "We are excited to provide our members a unique chance to compete to take the ultimate trip to space and to offer TV viewers and Web users a compelling online destination to participate in this fascinating reality drama."

Association with NBC and Mark Burnett was not a bad coup for us. It resulted in huge amounts of favorable press, with even a few die-hard cynics I knew rethinking their notions of space as a frontier for academics only.

To my surprise, though I suppose I should no longer have been surprised, Chirinjeev and I were raked over the coals by Legostaev and Derechin when they reviewed the final contract. By now I understood pretty well what could and should be done regarding the space assets, so there was little problem on the technical issues. But Energia was angry how little rights MirCorp had been given. We could not use the name *Destination Mir* for ourselves. We could not produce any merchandise for ourselves. We could not participate in the show without Burnett's agreement. We could not produce another game show or reality show for a period of time after the ending of the NBC agreement. From the Russian perspective, since it was "our" space program, we were in the driver's seat. In defense we had the hottest producer in reality television lending his name to MirCorp. This during a time when Russia had a poor image in the public's eye, and MirCorp was regarded as, to say the least, an irritant and threat to NASA. Only a producer with the hot hand of Burnett could have landed a major network so quickly. And in the face of some public skepticism and fear. A rival reality producer summed up that fear when he sarcastically voiced what many were thinking. "I hope they don't kill anybody."

Andrew Eddy sent me the results of an MSNBC poll taken after the announcement that seemed to reflect the public's fear of space travel:

What do you think of NBC's proposed "*Survivor*"-style show that would take everyday Americans to the Russian space station Mir?

-Its great! Everyone will want to watch that. 34%

-It's too dangerous; these "reality's shows are going too far. 66%

Conrad Riggs, with a quick smirk, commented how a blown rocket launch and killed contestant would sure get us all some fantastic publicity. I bet it would.

With the NBC deal we had a platform that would expose the Russian space program to tens of millions of Americans. We had an agreement that Burnett would also produce an hour long documentary on the Tito mission. We had, with great, great difficulty, worked out a waiver of Cameron. Conrad Riggs had insisted that the NBC game show would be the first media event involving the Mir. Tito was not as difficult an issue as Cameron, but finally there was an understanding that the markets were in no way competitive.

A board meeting with Energia in late August reviewed the MirCorp situation. From my perspective, the team had done the best anyone could have expected. During the meeting, it was agreed that Tito would fly in March and Cameron, if funds were identified, on a two month mission in the middle of 2001. But one small problem remained: there were no Soyuz's. No rockets were ready, and Anderson was late with funds promised for September.

In late September Semenov informed his government that MirCorp had missed a payment. Immediately an angry email from Rick Tumlinson popped into my inbox. Tumlinson wrote that had an American CEO undercut his company like Semenov, he would be fired. I was forced to disagree. In my view Semenov had taken exactly the right steps both legally and ethically. The Mir was drifting lower yet again and an upcoming General Designers Council meeting had to decide whether the launch-ready Progress cargo ship should lift the space station or prepare it for de-orbit. We were late with funds, not for this year, but for next. Yet again we were in a race against the solar flares, the international politics and the need for millions for the production line. I was invited to write a statement to be read at the meeting of the General Designers planned for October 3rd. In my note I explained NBC had already broadcast a commercial plugging *Destination Mir* during the closing ceremonies of the 2000 Summer Olympics. The show was real as far as NBC was concerned. I explained the Tito contract and its value in terms of branding, and all else we had accomplished in 2000. I concluded by asking them to launch the Progress on a mission to boost the Mir. For whatever reason, they did. We had bought ourselves another few months to solve the dilemma of the production line.

At the Apogee

The Autumn of 2000 was the apogee of MirCorp. Soon the incessant demands placed on Energia of having to supply ISS with Soyuz rockets without payments from either the Russian or American governments, coupled with the necessary Progress and Soyuz vehicles for the Mir, proved too much for Energia and an impatient Russian government. It all began to slip away from this moment onwards.

But let's stop here, in October of 2000 and take stock of the situation.

I had met with Walt Anderson and Rick Tumlinson at the Space Frontier Foundation Los Angeles event one year before. The MirCorp lease had been signed with RSC Energia in February, just ten months ago. The Progress boosting the empty space station was launched at the start of February. The Soyuz TM-30 MirCorp crew had docked with the space station in April. Legally, that docking gave us the contractual rights to the space station resources. Since its inception MirCorp funding had been used to keep open the production lines and helped Energia meet its hardware obligations. Four months ago we had signed with Dennis Tito. His funds for the ride would be used for one and possibly two new Soyuz rockets. We had just signed a stupendous media deal with NBC and Mark Burnett. We had an agreement with one ISS space agency. We were still working with James Cameron. We had interest from the Internet community to use Mir as a portal, and had opened negotiations at the highest levels with the Murdoch family and Fox. Negotiations with the Japanese aerospace company Mitsubishi were ongoing and very mature. If we could guarantee a multi-year Mir program, the Japanese would support MirCorp.

So far this year we had identified about $100 million in anticipated revenue. We had locked in several multi-year opportunities and handed to RSC Energia over $30 million in cash. The business model, the concept of a low-cost operator of a space station commercially marketing goods and services, had without a doubt been proven. The idea of the Mir space station being used at a fraction of the cost of the multi-billion dollar ISS, by all segments of the marketplace, from scientists and researchers to everyday people, had also been proven. Most importantly perhaps looking to the future, MirCorp was the first hybrid manned venture, bringing together a proven hardware manufacturer with a management focused on marketing, branding and sales. As such, we mirrored the passenger aviation industry, where two manufacturers, Boeing and Airbus, supply hardware to companies skilled only at selling tickets and building up the market. Government was certainly needed, as in aviation to provide safety and provide infrastructure to meet its own requirements. But, government was no longer necessary as an operator in low-earth orbit, just as the government builds the airports, runs the air traffic network, monitored the planes but didn't fly the passengers. What has proven so successful in aviation would, I was coming to believe, prove equally positive for the commercial manned space market.

What's more, our business plan had been executed in the face of an organized boycott by NASA, and in spite of the State Department's block on the export license for the tether. Not to mention a sophisticated disinformation campaign being conducted personally by the head of NASA.

From this plateau in October of 2000 we began our slow descent to earth, as would the Mir space station. It took a year for both to fall to earth and certainly there was still the chance to save the situation. But the path became narrower and more confining. After the Semenov notification of Anderson being late on the scheduled payments, the critics came out in force. The target of their wrath was MirCorp and its customers. First to suffer was Jim Cameron. Rosaviakosmos unleashed a public campaign against Cameron's plans to film aboard the Mir. Knowing he had long insisted on anonymity, unnamed Rosaviakosmos officials leaked Cameron's plans, and his medical testing, to the media.

Cameron was furious and dashed off an email to me saying that what the Space Agency was doing was "dirty pool." He added, "it may negatively impact my ability to raise funds with certain entities, but we can overcome this problem." Cameron went on to write that though "this does not dissuade me from planning to fly to Mir. I would request that they respect a future client, and one who will have significant media reach to comment positively on the Russian space program, by keeping our business private in the future. "

But the situation only grew more confused and cloudy from this point onwards.

TV's Race to Space

By late autumn of 2000 it was clear that Legostaev and Derechin were stalling on signing the NBC deal with Mark Burnett. Petty issues were raised, instead of focusing on the fantastic opportunity at hand. Something was amiss, and by December we understood. The German television production company Brainpool announced a deal with the European space company Astrium to produce a series called *Space Commander*. Seven winners, one from each season, will be rocketed to the International Space Station between 2002 and 2008. The German company had supposedly paid about $7 million to Energia for the rights.

The reaction from NBC, from Conrad Riggs, from Mark Burnett, from Walt, from me, was not a pleasing one. From the MirCorp perspective, our partner had undercut us. NBC publicly stayed loyal to Mark Burnett, but the trust towards us was now ticking away. Were the Russians supportive of our deal or not? Entertainment reporters had a good time, calling it a new generation space race, this time between the Germans, the Americans and the Russians, all fighting to be first with a reality show. Nationalistic pride set in immediately, with one German reporter proudly asserting that while "Germany may not have been a strong contender in the last space race, German TV doesn't intend to be left behind in this one."

In a stormy meeting with Legostaev, I learned that Energia had cut their "media rights" in half, one for the Mir and one for ISS. In their view, they had sold us the media rights for the Mir, and were now doing the same for ISS. The contract agreed with NBC promised Mark Burnett an exclusive to the Mir-there was no mention of the International Space Station. So Energia insisted they had been honorable in their dealings. There was no understanding that this negotiating duplicity might end Mark Burnett's interest. No understanding of how the entertainment industry worked. But it got worse. Right after my encounter with Legostaev I arranged to meet with Deep Throatskii in our usual spot at the lobby bar of the National Hotel, just across from Red Square. I was now meeting with this government official in 2000 more than over the past six years. From Deep Throatskii I learned a startling fact. That RSC Energia had also sold to Astrium the media rights for the commercial passengers on the Soyuz, or what we would call the "third seat." Astrium would later confirm this information.

Did that include media rights for the Tito mission? Astrium said yes. Legostaev said no.

Did that include Cameron's media plans? Astrium said yes. Legostaev said no.

And what about our back up plan to send the NBC winner up in our own Soyuz if the Mir was to be deorbited?

A few weeks earlier Walt had sent another multi-million dollar payment to Energia as a sign of his continued interest, despite his own worsening personal financial position. It was done to help assure that at least one of our deals would be implemented. "What the hell was the point," screamed Anderson, "if we don't own the media rights?" By the start of 2001, the situation with our media business just got worse. Burnett elected to push his own deal to the International Space Station and compete against Astrium. In an interview with the television reviewer for *The New York Times*, the columnist reported that "when word came last month that the aging Russian space station Mir might come down from its anchor high above the earth long before Mr. Burnett could realize his next big plan for reality television, a contest to take a trip to Mir, he was far from fazed. "We'll have a show," Mr. Burnett said, "I'm sure of that." Mr. Burnett, speaking by satellite phone from a remote section of the Australian outback, where he was supervising the second edition of *Survivor*, said he suspected some subtle involvement of NASA in the decision. "Still," he said if there is no space station available I'll just rename it *Destination Space* and we'll send someone to go around the earth 50 times."

But Energia's selling of the Soyuz media rights carte blanche clouded all such plans. Everyone was confused. Energia, suddenly panicking it would lose the lucrative entertainment market, then reversed course and asked us to compete in a competition for media rights to ISS, against Astrium and Mark Burnett. It was a highly unprofessional situation, with different Russian space organizations and companies competing against one another. Running out of funds we did a small round of financing, and businessman and friend Brian Streidel led the round. The funds provided us with the means to compete for the exclusive media rights. MirCorp won the media competition, but we all lost in the marketplace. In the end the networks walked, the producers walked, and a space-themed game show, or space-based entertainment show has yet to be produced.

Rosaviakosmos then made an attempt to sign with another American television network without MirCorp. I was told that Koptev and his advisors believed we had blown the NBC deal and they could do it better. But the differences between Hollywood and the Russian government officials were just too great, and it came to naught. I understood the showstoppers apparent from both sides, but could only watch from a distance as the entertainment market crumbled away.

I'm not sure what Energia was doing in giving away media rights will little understanding of the consequences. Perhaps it was an act of desperation, of realizing that Anderson could not pay for the next several Soyuz's and management believed it could split hairs by dividing up the rights to the Mir and the ISS. What was perhaps contractually correct from a narrow legal view would, and did, destroy the interest of the major media players towards working with Russia. There was no understanding from Energia that the entertainment business is above all else based on personalities. Directors and producers work with those they know and trust. And that there is no shortage of great projects, allowing these highly paid creative business people to quickly move on to the next project if uncomfortable. The entire episode left a bad taste with some very powerful people in Los Angeles. Hollywood may well turn out to be a major player in some of the exciting commercial projects now being planned, whether it is commercial space stations or private lunar ventures. But the space company providing the hardware, whether American, Russian, Chinese, Indian or European, will need to accept that much of the revenue comes from the "back end", as well as from merchandising, and that the producers and directors need very unique requirements in order to recoup their own upfront capital investments.

But the miscommunication worked, as usual, in both directions. The perfect example of this came during the unsuccessful effort to launch the pop star Lance Bass, whose business model rested on the revenue coming solely from sponsors and a television show. His producer David Krieff had signed with MTV, but the network pulled out, so they said to me, after seeing the raw footage of Bass, and didn't think the show would be a grabber.

There were critical problems attracting sponsors for the Bass project. As one major advertiser explained, "I can sponsor a boy band tour for $10 million and not worry that the idol of 30 million teenage girls will be killed on the launch pad." What was not disruptive to a game show producer like Conrad Riggs were to those being asked to support a pop star launch. Just when it seemed hopeless, Krieff signed on a television network to fund the proposed space show, but subject to one tiny contractual issue. The Soyuz carrying Bass had to launch on one particular day. But not just any day. It had to be on Halloween. The Russians were being asked to contractually promise to launch Lance Bass on the American holiday where children and drunken adults dress up in macabre dress. When we refused to sign, Krieff tried humor to change our position. "Listen, Lance has already chosen his costume, and you've agreed. He's spending Halloween as a spaceman." The Russians were not amused.

As Walt predicted, the huge time and effort with the world of television and Hollywood eventually came to naught, but not for the reasons he believed. Had we the investor funds we might very well have realized the NBC deal, which would have changed the entire financial situation. Still, the management of Star City, Rosaviakosmos and Energia never fully understood the unique business model and personalities from the world of entertainment. It's a huge cultural clash waiting to be overcome.

Chapter 31: 2001, Our Space Odyssey

James Cameron and Mark Burnett and John Daly and all the other Hollywood producers missed a factual tale far more outrageous than any science fiction story, and certainly more sinister than a reality show. The tale that could have been brought to the silver screen would begin when an empire called the Soviet Union collapsed, leaving behind only an aging space station as a symbol of its one time equality with America. With Russian government funds running low, this crumbling space station, called "Peace", was slated for destruction, in order to make way for the extravagantly expensive station supported by a cabal of American aerospace contractors. The new American station was called Freedom as a final jab against the disappearing Communist regime.

All seemed fine for NASA. "Peace" was weeks away from being

Energia maintained the production of space station launch vehicles during the collapse of the Soviet Union, the financial depression of the Russian Federation, the second grounding of the NASA shuttle fleet and the inability of the Russian government to ever pay more than half the costs. Under Yuri Semenov and operational director Nikolai Zelenschikov all flights were safely launched and all commitments kept. It is an accomplishment that has yet to be acknowledged appropriately.

Photo: Jeffrey Manber

destroyed when suddenly a private American citizen plunked down millions in cash to save the space station. Walt Anderson drew his wealth from a mysterious off-shore venture-capital firm called Gold & Appel, rumored to be named in honor of the second volume of the science-fantasy trilogy *The Illuminatus! Trilogy*. Here the story would go Gothic. Published in 1975, *The Illuminatus! Trilogy* weaves a mind-twisting trek through myriad conspiracy theories involving the likes of Adolf Hitler, the Kennedy's and Martin Luther King, together with a plot by a rock band called the American Medical Association to destroy the world.

In the second volume, entitled *The Golden Apple*, two New York City detectives seek to liberate society from those in power. *The Golden Apple* is a symbol referring to the discord present in the universe. Presumably, the reclusive Walt Anderson is sympathetic to the anarchistic slant of the trilogy. His giving of a golden apple to the CEO of MirCorp, his company created to save "Peace", leaves little doubt his emotional connection with the trilogy. In this true to life space tale, the American government official sent to stamp out Gold & Appel was named Dan Goldin. As the head of NASA, this "Golden Apple Slayer" took aggressive actions to destroy the rebel threat in order to preserve the status quo.

Was Goldin acting on his own, or at the behest of the powerful aerospace companies? Who was leading whom? And, in a further twist, the libertarian Anderson joined forces with a Russian space official turned free marketer named Yuri Semenov, who was already the number one enemy of Dan Goldin.

As the battle entered its second year Goldin-with his army of contractors-was again poised for victory. The libertarian investor had been hurt by the March 2000 market crash and his funds were perilously low. But, another unexpected, random occurrence takes place. Against all odds, a California businessman named Dennis Tito was willing to buy a ticket to ride to the space station. If the voyage took place, the funds would save the space station for at least another year. If the Tito mission could be thwarted, the unwanted station would be doomed. To provide a little more pizzazz, the producers of the film would place the climax in the year 2001. This was the year in which Arthur C. Clarke chose to base his classic thriller *2001, A Space Odyssey*. The circle between fact and fiction was made complete by discovering that Walt Anderson and his entourage flew on Anderson's private jet to Sri Lanka, telling a delighted Clarke of the upcoming battle with NASA. The science fiction writer was sympathetic, noting how the routine space travel envisioned by 2001 was nowhere near reality.

How would our improbable story end? Who would be the victor? And perhaps most important, what lessons can be learned from this real-life space odyssey?

Dennis Tito never sought to be a hero in the clash between two differing paths of space exploration. The veteran financial executive was politically conservative, an ex-Marine and former NASA employee who sought only to realize his life-long dream to experience space travel. He was also true to his word. When Tito told me he would salute and obey, this is exactly what took place. As promised, the $1 million down payment was wired to MirCorp. When told of a medical problem that needed to be corrected, he underwent corrective surgery. Problems that did arise were discussed, and common sense solutions were sought. Chief among the early issues was just how Tito would train for his mission. It was my desire to reduce dramatically the training time necessary for any private traveler. A solid core in Russia, including some of the Energia cosmonauts, insisted that anyone entering the Soyuz should speak Russian, should train with the crew, should know how to survive in any emergency and should, in general, be considered a full member of the crew, complete with the customary years of training.

Tito wanted to spend much of his early time in a hotel in downtown Moscow, rather than at Star City. He wanted to have open communication with his LA office. This provided me with the excuse to try and trim requirements for paying passengers. It was a long and difficult fight, putting me at odds with senior officials like Ryumin. In the end, there was some compromise and the required training was reduced, but even Tito began to side with Star City. "These guys are the experts. If they feel I need to bond with the crew, then I'll bond." Some of the inflexible traditions were amusing. When granted his papers formalizing his status as a cosmonaut in training, the businessman was awarded with a document card granting him free travel on the Moscow metro. Tito really never learned to speak Russian, but did spend mind-numbing hours learning intricate and unnecessary details of the Soyuz. In the end a face-saving compromise was reached. While Tito trained in two week spurts, the overall time would be counted as his training period. So the Russians could say Tito trained for an eight month period, yet in reality his training occupied only a percentage of that time.

While grappling with the Tito issues, we struggled to find the means to survive. In October of 2000, the prime minister of Russia announced it was time to bring down the Mir. This public announcement again sent NBC reeling, not to mention the potential Japanese investors. It also scuttled a deal with a major Internet portal. But others in the Russian cabinet disagreed, and quietly I met with one cabinet official and several Kremlin advisors to explain the situation. Who was behind arranging these meetings? I wasn't sure. There were ties to St. Petersburg, home of Vladimir Putin, and ties into Energia. But the motivation of my helpers, and their own reasons for helping MirCorp, was unclear. The Kremlin aides were serious younger men working in stately offices across the Moscow River from the Kremlin. It was the new Russia. No bumbling government officials drinking far too much as in the Yeltsin era. These men were technocrats, carefully analyzing each political step. As always in Russia, exactly who would benefit if MirCorp remained in control was not perfectly understood, though I gained some shadowy answers. Deep Throatskii was concerned at the rarefied atmosphere that I had entered. "You are a foreigner," he nervously reminded me. "A friend, yes. But a foreigner entering the realm of a very complicated dispute." What would he advise me to do?

Guarantee nothing that is not rock certain was his careful answer. "Oh, and don't lose Semenov's support for you personally. He is your protection." I obeyed his advice to the letter. Chirinjeev wanted me to promise that another ten million in funding would soon arrive. Derechin and I sought a bank guaranteed note, but to no avail. I reported the exact situation, however disappointing, to each of the Kremlin advisors. Some ominous signals then began to appear. A long article was published in a Russian business journal, for example, about MirCorp. The reporter stated that the source of "my power over the Russians" was due to my having a relationship with the daughter of Yuri Koptev. For centuries Russian nationalists have spoken darkly of foreigners with undue powers over them. It was not a good sign.

Two Soyuz spacecraft were urgently needed, one for ISS and one for the Mir. Ours was needed first. Everything was taking too long and becoming too costly. The establishment of the Tito escrow was a time-consuming process, involving plenty of lawyers. We brought in a great financial officer, Helena Hardman, who sought to structure the escrow in the best light possible. That meant, as it turned out, that MirCorp would be paid when certain program milestones were met. One of our remaining hopes was that Tito would see fit to help us. Zelenschikov was standing ready for three work shifts of eight hours each to work around the clock to produce the Tito Soyuz if the funds were there. Or if Tito would take some now illiquid Internet stock from Anderson as collateral and release the funds, we could also survive.

Tito saw the situation differently. He cared little for the philosophical aspects of our battle that at times seemed religious in tone. Several news stories picked up on this theme, describing our survival into the new year as the "MirCorp Miracle". Yet for Tito, his was a personal mission, not one to make a statement or promote commercial space. Tito wanted to fly. He was enthralled by the prospect of the publicity, the media programs, the interviews, but above all else, his quest was to experience spaceflight. If MirCorp was the platform, fine. If Energia was the platform, fine. If NASA would have allowed space tourism on the space shuttle, even better.

In October of 2000 we floated the concept of a MirCorp IPO, meaning to sell shares to the general public. Despite the poor stock market conditions and huge political uncertainties, the financial community was receptive. *The Financial Times* ran a story which brought enquiries from the London financial community. A frantic round of meetings in London and New York suggested that if we had NBC and Mark Burnett, and we had control of the Mir space station, one could envision an IPO allowing us to raise about $50 million. Far less than our initial plans, which was to fly Tito and then, prior to the NBC deal, float a percentage of the company for several hundreds of millions. Yet again, like so much of our project, events worked against us. The Wall Street brokerage firms and British private hedge funds required hefty upfront fees to plan the public offering. All promised we could have something on the "Street" by the end of 2001. That was far too long. Worse, Koptev publicly poured cold water on our plans. Andrew Eddy left at this point, not comfortable with the IPO plans and probably just as weary as all of us.

It was a time of frustration. I was imploring our two investors to provide a simple yes or no whether new funds were forthcoming. While keeping the staff and office funded, they sought to find a creative solution, whether an advance from NBC or borrowing against Tito's escrow, or the IPO. But NBC's lawyers were insisting the Russian government agree in writing to keep Mir flying, as well as Energia had to explain, also in writing, the media situation involving Astrium. Derechin of course was balking. Nothing seemed promising. Anderson would sit with his pages of stock printouts, some with thick black lines crossing out a company name, as CEO's of all his ventures were madly demanding capital for each of their projects. Some needed a million. Some needed a hundred thousand dollars. And there I was, with Anderson in Amsterdam, while he tried to squeeze ten million from his now illiquid holdings. In a dramatic and emotional meeting with Semenov, Anderson managed to wire another $4 million. Semenov was profoundly grateful, and looking Anderson straight in the eye, and his hand over his heart, the General Director promised that if the Soyuz manufactured with these funds was not used for MirCorp, the funds would be returned. As the meeting broke up an agitated Derechin rushed up to us and blurted that "Dr. Semenov had misspoken about returning the funds." And thus was created a fatal ridge between the two men. It was the last payment from Anderson.

During this time MirCorp was both the path to realize Tito's dream and the obstacle. It was an uneasy situation. In late 2000 different Russian organizations, including Star City, began fighting to take control of Tito's mission and hence his funds, while Russian space agency officials were publicly putting down MirCorp. Even as we moved forward with Mark Burnett and NBC in due diligence, Energia was growing increasingly desperate. The production line was idle. The industry boycott remained in place. The only identified money now "in the bank" was from Tito. In a November meeting held without us, someone from Energia promised Tito he could fly to the International Space Station if the Mir was deorbited. And that was the end. The prospect of being first to the ISS was too great a thrill to turn down. Rumors flew first through the Russian community and then conveniently leaked into the newspapers.

Soon after electing to shift to ISS, Tito requested to meet with me at Star City. During the meeting he asked if I was willing to help him in the transition to ISS. Both Chirinjeev and Walt were adamant in their refusal, believing there was still hope to save the Mir. The company Space Adventures was fighting to take control of the Tito program as well, backed by some powerful Russian officials. Both Energia and Tito suggested we work together, but Anderson refused. Anderson's stubbornness was disappointing, but not for business considerations alone. It had more to do with my squaring it with Dennis.

Tito had been angry with me after I briefly toyed with finding another candidate to fly first with us. He remained angry as Kathuria was often his link to MirCorp, and not me. I wanted to show him that there was far more at stake than a contract, and I would fight for his right to fly; NASA be damned. There was also another reason. I had worked too hard and for too long to let the Tito situation end in confusion. Clearly space tourism was on the verge of reality and though I didn't believe sending tourists in a government vehicle to a multinational governmental space station was the future, it was at least the first step. Eric Anderson, the president of Space Adventures, shared this perspective, despite his firms' success in brokering rides via Rosaviakosmos. Speaking to *Cosmos Magazine* in 2004, Anderson admitted "That is not space tourism," referring to the Soyuz missions with paying customers. "That is just something for billionaires to enjoy." What Anderson meant was that a more commercial market would be to have a truly non-government infrastructure for the vehicles and for the space stations. That day will have to wait, but the first step was to get Tito into space.

At the time, Derechin and I were under no illusions how difficult the battle would be to send Tito to what was effectively the NASA space station. Derechin asked me privately to help and I agreed. Tito had the situation half right when he told a reporter "I don't believe the Russians are eager to default on my contract. There is a considerable amount of money involved, and the Russians are much more savvy these days about world opinion." Fortunately for Tito, the Russians did not care one bit about world opinion.

In January of 2001 the prime minister signed a decree authorizing the State to take back control of the Mir. In one last attempt, it was arranged that a letter written by me would be delivered to President Vladimir Putin. We also published the letter in the leading business newspaper in Moscow. But the public letter and the private were different. One could not say publicly what I saw as the reasons to continue the Mir throughout 2001. In my public letter I wrote that: "It is my honor but none the less unpleasant task to write to you in the hopes we can together find a solution to keep in space the Russian space station Mir." I explained that "We believe that Energia is being asked to act as both a company and rely on its own, and as a government agency, providing 17 launches to keep ISS in orbit, and for what purpose...to allow NASA to move ahead in this important technological industry?"

In the letter I asked for time to find international funds for supporting the Soyuz and Progress production lines, upon which not only the Mir but also the ISS was dependent. In my private letter, language was included that had been suggested by the shadowy Kremlin advisors. My letter was presented to Putin at the next cabinet meeting. But the stars were moving away from us and there were now cross-purposes even within our partner Energia. From the perspective of some at Energia, Tito's funds for an ISS Soyuz would allow them to honor their international obligations and extract Semenov from the increasingly bitter internal battle for control of the Mir, now a symbol not of scientific might, but of Russia's fading international influence.

Walt visited Tito and tried one more time to free up his funds or to borrow against them. "You have to understand," Tito brutally explained, "I don't give a damn about MirCorp or the Mir if I can fly to the ISS." Nor did I give up, arguing to Tito that without MirCorp he would lose both the media platform and the branding. I'm not sure whether he didn't care or didn't believe me, but I turned out to be correct. The lack of an underlying positive story caused producers from the BBC and American networks to yet again pull back, and the Tito mission ultimately launched without any documentary or television platform. But in our concern for media coverage we did lose sight that for Dennis Tito our crusade was not his. That's how it is in the most heated moments of the battle. There is the tendency to believe that those fighting on the same side share the same goals and objectives. Energia needed capital for the production line. Anderson and Tumlinson wanted to disrupt the status quo. Semenov wanted the capability to tap private international markets for funding. Tito wanted to fly. Cameron wanted to do a movie. For a time, our interests coincided and so we all fought together. But soon MirCorp stood alone. Semenov wanted to continue on a truly commercial path, but not risk failing to meet Russian promises to NASA.

However, the General Director did not give up on the dream of a commercial space platform. During 2001 we began to discuss the idea of a MirCorp free-flyer, a small man-tended platform that could house tourists and conduct commercial research without NASA. The economics made sense. If we could send two tourists a year, plus a reality show, it would have been economically feasible. In our plan the Soyuz would go first to our small station and then to ISS, to which NASA responded favorably. In his own way, Semenov was gently pushing Anderson and MirCorp away from the Mir while respecting Anderson's loathing of ISS.

Chapter 32: Into the Watery Grave

March 23rd 2001

The last day in the life of the Mir space station was sadder than I had expected. As the Mir came streaking back to the earth, I found little pleasure in watching the television reports or reading the dozens of emails from well-meaning friends. The station touched the earth's atmosphere high above the skies of Australia before obediently settling into its watery grave. Semenov's engineers had done this final job with their expected professionalism.

CNN's Miles O'Brien had sophisticated computer simulations allowing viewers to better visualize the final moments. But ignored was the fact that the space station was notable not only as a piece of hardware, but also as a home. Not shown were snippets from the dozens of home movies that existed of life on the station. The men singing their folk songs, laughing over the well-worn jokes, playing chess, or watching in zero-gravity some

The Mir space station. February 20, 1986-March 23rd, 2001.

Photo: RSC Energia

favorite films with the appreciated plastic packets of French wine or Russian vodka. Or greeting new arrivals with the traditional Russian bread and salt. Mir was filled with bric-a-brac from the more than one hundred men and women who fought through the depression and fear, life-threatening dangers and heart-pounding exhilaration common in any frontier. In my mind's eye I could see the more than 135 tons of modules and scientific payloads and guitars and photos of loved ones and national flags of different countries and leftover food and oxygen, spare boots and gloves, Alexander Polishchuk's cosmic sculpture and a few dozen movies burning in the hot hell of re-entry.

In America we derided the Station from start to finish. When new it was a tin can, nothing like our soon-to-be Freedom. When in its prime we fearfully boycotted using it as a genuine tool for learning and heaped scorn on the commercial underpinnings. When finally given the chance to live aboard the Mir, few in the space agency took the time to understand what the Russians had perfected both in terms of hardware and human training, and in its decline we laughed openly. Always, for the American establishment, Mir was an irritant.

Lost on us was that the designers had set out to build more than another in a series of Soviet space stations. If their goal were solely technical- a newer space station- it would have been named Salyut 8. Instead, it took on a new name. "Mir" is translated as being Russian for "peace," but it is one of the more complex words in a complex language. Boris Artemov, way back in 1992, explained how the word also refers to a peasant village, where the locals harvested their own crops, and kept the proceeds, without interference from the local representatives of the Czar. It was a self-contained collective, just as Semenov always referred to Energia. The word over time also took on connections to the Orthodox Church. The Czar was believed to have been "Miropomazanik Bozhiy," or chosen by God. Boris went on like this for quite some time, writing down different verbs and different spiritual expressions based on using "Mir" as the root. I wondered aloud as to the political point these hidden, secret space designers had in naming the station. Was it a conspiracy against the Kremlin leadership?

Artemov just stared at me, not answering. He had revealed enough, I suppose. While watching CNN track the disintegrating space station, I silently saluted those unknown engineers for having achieved their apparent objective of creating a commercial outpost run by the "collective," with muted interference from the Kremlin.

In Russia the outcry was loud. The inevitable end of an era of Russian domination in manned space activities, proclaimed the pundits. Yuri Koptev struck back, calling the re-entry a technical success because of international cooperation. The space experts in TsUP worked in real-time with the United States Space Command, buried deep in the Cheyenne Mountains. U.S. Secretary of State Colin Powell cited this cooperation as proof of good relations between the two countries, even as America and Russia were each expelling some 40 diplomats that seemed more like the spy games from the Cold War. Powell was aware of these echoes and described the expulsions as "a stand-alone problem we had to deal with" and cautioned against viewing the developments as putting relations in a deep freeze. "Even as the incident unfolded, U.S. Space Command officials were working with Russian technicians to chart the path of the space station, "noted the Secretary of State.

Yes, muttered the Russians, true cooperation to make sure our station gets buried.

While the U.S. Secretary of State and the director of Rosaviakosmos publicly considered the destruction of the Russian space station as a wonderful moment in superpower cooperation, the event was viewed differently in the Duma. A bill introduced on the day after the de-orbiting called for Koptev's head to roll. It concluded by stating: "The State Duma deems it expedient to relieve Mr. Yuri Koptev from his post of the head of the Russian Aerospace Agency on the grounds that he erroneously informed the President and the Russian government of the available potential to continue the operation of the Mir.

"Mr. Koptev bear (sic) personal responsibility for the Mir's premature liquidation, and also for increasingly drawing Russia into the costly and technically unjustified ISS project." The Duma was certainly reflecting the public sentiment in Russia. Not just regret at the loss of the Russian space station, but a belief that their involvement in the ISS was not best for the interests of Russia. Of the almost one hundred e-mails we received from Russia, one stands out as unfortunately typical. It was a terse one-line message.

"Americans will long savor the bringing down of Mir."

The patriotic Russian writer was wrong. Because Americans would within a short period of time move on. Like so many battles fought on the political and military battlefields it is the losers that have the longest memories.

American observers also had a lot to say about the termination of the space station. NASA, despite the supportive front put on by Powell, was cited as the culprit by many. Ed Hudgins at the Cato Institute in Washington captured the situation with a report that began, "The Mir space station was born as a symbol of communist strength and died as a noble experiment in private space commerce. NASA officials helped kill it. It's odyssey provides lessons concerning why so few individuals live and work in space today, and what must be done if space is to truly become a domain of human activity in the future."

For other reporters, the Mir remained a stereotype of the collapsed Soviet empire, as emphasized by one television reporter. "It was a disaster, nothing more than a joke, a deathtrap for our astronauts."

"Let's see how the International Space Station looks in 15 years," retorted reporter Brian Burrough, author of *Dragonfly*, a book on the NASA astronauts who lived onboard the Mir.

Joel Achenbach writing in the *Washington Post* combined the two views. After describing the space station as an "malodorous, filthy…orbiting dumpster. It caught on fire. It got moldy," he went on to describe the Mir as one of the great characters in the still-brief narrative of human spaceflight. "The defenders of Mir are probably right when they say it had just one bad year, 1997."

Achenbach then adroitly switched from Mir-bashing to the issue of just why the Mir was destroyed. "Modesty and pluck are no longer in fashion in Low Earth Orbit," wrote Achenbach. "Now we have the International Space Station, a grandiose money pit. The Mir seems sensible by contrast. It didn't exist to funnel money to aerospace corporations. …The ISS has been planned, in one form or another, for decades. Five times the size of Mir, it will cost roughly $60 billion. In a sense, we had no choice but to build it--because the Soviets had Mir. We couldn't let the Russians hold the high ground. Think about that when you read about the Mir's demise today. It will be gone, but we'll still be paying for it."

For his part, Semenov was unsentimental. In an interview in the *Russian Gazette*, Semenov dismissed unfavorable comparisons by his cosmonauts between the Mir and the new ISS as "nostalgia." Semenov took credit for the International Space Station, saying how "the very idea to build ISS belongs to RSC Energia. This idea was born as far back as 1993," when according to the General Director, the flaws in NASA's approach were clear to his experts, and a proposal by Koptev and himself to build a space station based on the Mir concept was gratefully accepted by the United States.

Few at NASA realized that Semenov's experience in managing space stations began in 1969, when he was given control for the development of the Salyut space stations. He oversaw increasingly complex stations that finally resulted in the Mir and, one can argue, the International Space Station. He had seen each of the space stations end their lives, only to be replaced by an outpost larger and more sophisticated. With unmatched experience, the General Director explained with surgical precision how the Mir should be remembered. The Mir "laid a foundation for the classic approach to space station design: principle of modularity, multidisciplinary scientific experiments, adaptability to different tasks, transportation support, and many other things. All that we first tried on Mir, we are now going to apply to the international station on a new level."

There were other notable innovations resulting from Semenov's Mir, which the hard-nosed head of RSC Energia omitted to mention. That his organization used the outpost to open the space frontier to everyday people. How Energia forced NASA to pay for visiting the space station, rather than allowing its use as a diplomatic tool. But above all else, Semenov endured ridicule for showing that by selling services for this village called Mir the free market has a place within space exploration. That is a legacy that a new generation of American space entrepreneurs have taken as their own starting point.

If that alone is the epitaph of the Mir, it will wear well in the centuries to come.

Chapter 33: The Final Confrontation

In the flick *Men in Black*, alien life forms fight among themselves on crowded New York streets while earthlings remain oblivious to the struggle. The Men in Black, out to save the earth, are never seen or known to everyday people. So too with the clash between MirCorp and NASA. People knew of our battle, but it wasn't high enough on the "radar screen" to really capture their attention and have them dig deeper. Until the Tito battle. Why Dan Goldin chose to go ballistic over Dennis Tito and elevate the dispute into a public battle is unknown. Some have suggested it was the offended pride from a government official who had become far too personally involved in the dispute with Semenov. Or that it was the genuine anger towards everyday capitalism besmirching the governmental International Space Station. A more mundane explanation could well be that in January of 2001 the new administration of George W. Bush replaced that

From 1992 onwards NASA Administrator Daniel Goldin (left) cherished the special relationship with his Russian counterpart Yuri Koptev. But during the public opposition by NASA to the flight of American businessman Dennis Tito, Goldin pushed too far, causing Koptev to stand firmly with Yuri Semenov in assuring the Tito flight took place. Photo: Bill Ingalls/NASA

of Bill Clinton. Goldin was asked to stay on. Soon he would be the longest serving head of NASA. The administrator's very public outrage against Tito might well have been an eager Democrat appeasing his new masters.

Whatever the motivation, NASA's intransigence towards the Tito mission was the final struggle in the decade of conflict between Yuri Semenov and Dan Goldin. NASA's stance against Tito also delivered for Walt Anderson the very public battle against NASA he had long dreamt about. In our final defeat, we achieved one of our most important successes.

For MirCorp, there was a lot of business to clean up and decisions to be made regarding our future. We held one final formal MirCorp meeting on Dennis Tito in February of 2001. In the General Designer's conference room the usual players on the Energia side lined up to meet with Walt Anderson and myself. There was Ryumin and Legostaev and Zelenschikov and Derechin. Semenov began by clearing his throat; the blustery boss was nervous. He explained that the Mir was indeed coming down and he wanted to know what we were prepared to do with Tito. Then he launched into a long explanation, formally thanking the American entrepreneur for his fight to save the Mir. "We will never forget what you have done. In this organization and in Russia, you have our thanks" the General Director solemnly promised Anderson.

Not one to ever stand on protocol, Anderson surprised the Russians by asking about his $4 million. After some hesitation and exchange of pretty frank language, it was agreed by Semenov that the funds would be moved to our next project, the small commercial space station. We then pushed hard to have Tito remain known as a MirCorp project, even though Space Adventures would handle some of his day to day logistics. Energia agreed, both in the board minutes and in a follow up letter. The flexibility by Energia management was probably due to a simple fact: we held Tito's money and contractually still controlled Tito. The choice of what to do was mine. It was slightly within my rights to delay returning the funds to Tito for subsequent payment to Rosaviakosmos. We had contractual issues with Energia and Star City, and I had long ago learned from the master himself, Yuri Semenov, that when you have a bit of control, you can dictate the situation. Hence his nervousness at that meeting. Surprisingly, Anderson agreed. "We need to realize something from the Tito situation, let's do nothing until we win."

But before the board meeting I took Derechin's old admonition to "know what we have signed," and I reread the agreement, the one signed by Semenov, Tito and myself. I reluctantly concluded that the best I could

THE FINAL CONFRONTATION

do for space tourism was to release the funds without preconditions. It was also necessary for me to sign a document, along with Semenov and Tito, which would release Tito from the Mir and send him to ISS. Much to Tito's profound appreciation, that's what I did. I signed an authorization note, and a wire was sent from our bank that was the mirror opposite of that order more than a year ago from Walt to his own bank. This time funds left our account and would be used not to rescue the Mir but to prepare a Soyuz for the government-controlled International Space Station. Anderson was angry, I think more at the overall situation than with me. But I didn't want anyone, anywhere, to blame a private company for screwing up the Tito mission. Mike Griffin's warning to me from years ago, about being the hero or goat, had not been forgotten. I didn't believe myself the goat, with everything that had been accomplished since opening Energia's Washington office. But the reality was different. Many industry figures were willing to blame those working with Russia for just about everything, from ISS problems, to harming American-Russian relations. This was part of my motivation. I wanted badly for Tito to realize his dream and fly.

Also lurking in my head was the warning from Deep Throatskii not to lose the protection of Semenov. This was uncharted waters and Russia was not the place to be without a local guide. With the funds returned to Tito, he was free to sign up for the April 2001 Soyuz taxi mission to the International Space Station. And that's when all hell broke loose. A full throttled conflict that pitted the obstinate administrator of NASA against an equally angry and united Russian space community. Whatever Goldin's motivation, he hurt NASA when electing to publicly draw a line in the sand against the Tito mission. This was a battle the American public could follow and support fell to Tito and even to the Russians.

The effort by NASA to block Tito's mission took place on all levels, operationally, politically and even psychologically. The extent of NASA opposition was made clear in February when I was called by the NASA liaison from the Bush White House, asking to speak when I was back in the States. A week later I returned the call. He wanted me to pass along a message to Tito. Choosing his words carefully, the political operative suggested that the White House would "appreciate" the Republican financial executive delaying his mission for at least six months, maybe "just a little longer." And, to put a little zing into the message, the White House liaison also warned that the White House would be contacting Los Angeles mayor Richard Riordan, whose Republican campaign Tito had strongly supported. The message I took away from the call was that Tito would not be welcome in the Republican party, from the White House to Los Angeles, if he flew to the ISS in April. Tito brushed off the White House threats.

At the request of the Russians I paid several visits to NASA Headquarters seeking to defuse the situation. I was told by John Schumacher that any discussion was moot. Tito would not fly. It was, he said, an issue of the highest concern and could impact American-Russian relations in the new Bush administration.

The first public skirmish took place on March 19th, when Tito arrived at the Johnson Space Center in Houston, Texas, to train with the Soyuz crew to gain familiarity with the American side of the space station. Now that the Russians considered Tito a crew member, his travel to and from America was being handled by the Energia Ltd. office, and thus coordinated by Kristin Oland. So we knew that Tito was traveling with a security team, an "embedded" reporter working on a Tito book and at least one assistant. In Houston, right at the front entrance to JSC, NASA manager Robert Cabana met the Tito group. Cabana dramatically announced that NASA would begin the training sessions with cosmonauts Talgat Musabayev and Yuri Baturin, but not with Tito. What came next was unique in space history. The cosmonauts, under orders from Semenov, refused to train and returned to their hotel, in solidarity to their crewmate. Soon enough the Russian cosmonauts began their JSC training. But the point had been made.

The standoff burst into the media and Americans just sort of scratched their heads, wondering what the fuss was all about. If a rich American wanted to spend millions to fly into space, why not? Reporters pounded NASA with the very logical question of what could be done if the Russians refused to listen to NASA. It was their Soyuz, was it not? Tito remained in Houston while some sort of solution was sort. At first, NASA refused to bend. Nor did the Russians, with Koptev and Semenov united in their determination, and their budgetary need, to implement Tito's contract.

With tensions running high, Tito returned one evening to his hotel in Clear Lake, alongside the Johnson Space Center, to find his hotel room had been ransacked. His bed was a mess, drawers were opened, his personal belongings scattered about the room. Tito flipped. He, or someone from his team, called Kristin and asked her to formally complain to NASA. The folks at JSC denied any involvement. Tito next used his own connections and reached the CEO of the hotel chain to complain. The incident seemed to have unleashed in Tito all the emotions he had carefully held bottled inside for the past year. The tension of working with the Russians. The collapse of the Mir program. The months of training. His embarrassing treatment at the hands of his own country. This violation of his privacy was the final straw. The situation was diffused when the head of his security detail came forward and admitted that they had messed up the room as a joke. If true, not a good joke; emotions were now rubbed pretty raw.

In a bureaucratic face-saving move, a committee led by former astronaut General Tom Stafford was created to suggest the ground rules by which non-professional candidates could be included on the ISS manifest. A very draconian series of recommendations came out, including calling on Tito to stay in the Russian side of the station. It also stipulated that Tito sign the standard space station agreement prohibiting astronauts from realizing any financial gain from being onboard the ISS. Part of the code reads as follows:

"ISS crewmembers shall refrain from any use of the position of ISS crewmember that is motivated, or has the appearance of being motivated, by private gain, including financial gain, for himself or herself or other persons or entities. Performance of ISS duties shall not be considered to be motivated by private gain."

Why? When a government employee, up to and including the president, finishes their public service, nothing prohibits "private gain," such as a book from being written. Why should a NASA space station astronaut or visitor be treated differently than holders of the Oval Office, or an Iraqi veteran, or a researcher at an Antarctic research facility? Imagine the outcry if the ISS regulations drawn up by NASA became the de factor law for all government employees. That perplexing question again reared up. Why is space different? It was as if NASA fears a fatal allergic reaction to personal gain.

Tito signed the agreement.

More was to come. NASA demanded that the Russians assume liability for any damages caused by Tito. Koptev was genuinely worried, explaining how "the NASA specialists presented us with a situation where our legal costs for Tito could be higher than his payment to us. Not only were we responsible for any possible damages, but also could be billed for crew time lost by NASA." As Derechin commented to me, "finally, your NASA has learned how to behave in a commercial manner. Pity it is inspired by such a stupid motivation." Energia and Rosaviakosmos were very concerned. The Russians, mindful of the quick-to-sue American legal mentality, were fearful at signing such an open-ended obligation. This dilemma was solved when Tito agreed that he would personally be responsible for any damages and lost crew time. No bill was ever sent by NASA.

With each new move by Goldin the Russians grew more united and more furious. As Derechin told a reporter, "They are just looking for phony excuses. However, they are going to learn very soon that we are absolutely determined to fulfill all terms of our contract with Tito in April. He signed all the paperwork, paid the money and we are going to do our part, no matter what they think." The fact that Derechin said this publicly was yet another sign of the anger felt by the Russians.

As common throughout his tenure, Goldin misplayed the Russian situation. The American side seemed not to care or understand that the disagreement over Tito was reaching a climax just as the Mir was being prepared for its late March destruction. All of Russia was united in the despair of being second-fiddle to NASA. Goldin may have thought he would continue to enjoy the support from Yuri Koptev, but the government official was under enormous pressure. Prime time television shows in Russia were dedicated to the coming demise of the Mir, and the coming era of, as it was seen by many, the subservient Russian role to the Americans. Rumors flew through Moscow that TsUP would be forced to close, that Star City would no longer be necessary. Goldin did nothing to assuage the fears, acting in a bellicose fashion as the end of Mir drew closer. On many of the local levels of NASA, it was a different story, as friendships had developed between Russian space officials and Americans from Houston and Huntsville and even NASA Headquarters. But at the top, there was little sign of how delicate was the political situation.

For Semenov, the NASA refusal to allow Tito to fly was the final insult. Energia had built the Mir space station that finally gave NASA long-duration experience. Energia and other Russian contractors had built the key modules that formed the ISS. Energia was paying out of its own pockets for a portion of the cargo and manned vehicles to the NASA station. And while his experts were readying the Progress mission to destroy his beloved Mir, NASA was denying him access to the only funds available for the production line. He believed it was politics, and dirty politics at that. "We have excellent and complete understanding with our US colleagues at the level of people who do the actual work. But," he added rather carefully, "at the level of politicians, other motives come into play." After the Tito mission Semenov shared with a reporter his thoughts. "We had a very serious funding problem. The government owed us more than a billion rubles. The guaranteed life of Soyuz TM-31 was expiring. So we proposed to under-

THE FINAL CONFRONTATION

take the replacement of the vehicle, by bringing to the station a paying tourist. We would earn through this a few millions, which, by the way, we wouldn't spend somewhere else, we would invest the funding in our joint ISS project. And we were insistent: either we fly with Tito, or we do not fly at all. RSC Energia never went back on its commitments."

Having met all of Goldin's requirements, Tito finally underwent his Johnson training. Still, the Russians were ready for some last minute maneuver by NASA. In a final stormy meeting at Energia just before we all left for the launch, Semenov was in his best form. His fist pounded on the table, his voice boomed loudly for all to hear. "We are leaving for the Tito mission. We do not return until he flies." The conference room, filled with Energia experts, broke into loud cries of "hurrahs."

Walt and I had been invited to attend and to my surprise, Anderson accepted. At first he sought to fly directly to Baikonur in his own plane, but the logistics were far too difficult. The only non-Soviet plane ever allowed at Baikonur was that of the Prime Minister of France. I'm not sure why Anderson attended this launch and not that of our own crew. But we did fly down. When we arrived on site all of us were to learn there was one more dramatic gamble by Dan Goldin to stop the Tito Soyuz mission. On April 25th, 2001 the Americans announced that the computers on the U.S. module Destiny had crashed. Efforts to reboot had failed, causing a delay to finishing the work of unloading the space shuttle Endeavour that was docked to the ISS. The Soyuz launch was set for Saturday. On the operational level, the NASA and Russian teams discussed the situation, with NASA asking for a launch delay. The news wires dutifully ran the story that the Tito mission would be delayed because of computer problems.

Semenov saw it differently. The General Designers had already met and approved the Saturday launch. The State Commission, which had the final say, had signed off. All was ready. Semenov's trust of NASA was so low that Semenov did not believe the space agency was being honest. He wasn't sure if this was one last move to get under the skin of the space tourist, like a time-out called just before an attempted last minute NFL field goal. Semenov long preached that space crews had to operate in a zone of confidence and calmness, leaving behind any earthly problems. Crews slated for multi-month space missions were helped in cleaning up their domestic affairs, and teams of experts made sure they were mentally prepared for leaving their earth-lives behind. No one was better than the Russians in achieving the right mental state for departing crews. Maybe he was reacting like a professional manager defending his prized rookie player, but

Semenov was not going to let NASA delay his mission, and possibly psych out Tito; not with the whole world watching.

Two nights before the launch there was an all-hands meeting at Baikonur with Semenov and Koptev presiding. Koptev wanted to delay. His voice somber, the tension showing on his reddish face, Koptev spoke slowly, laying out the reasons to respect NASA's request. What's a few days? If something went wrong, either with the docking, or with Tito onboard, the whole mess would now be his fault. But Semenov was not budging. It was the most emotional performance I'd seen from the boss. Screaming, his voice pitching to a high level, he had enough with NASA, with Goldin, with being told what to do and when to do it. *We are launching on time.* Forget NASA. These people are seeking to embarrass us. And if they are telling the truth, one computer is now working. *We are launching on time.* He ended his tirade with such suddenness I worried he had been stricken with heart trouble. He looked in pain. No one in the room dared say a word. Semenov began to say something more, but a very pissed off Koptev stopped him with a wave of his hand.

On Friday word went back to NASA that the Soyuz was going to launch with or without the computers functioning, and with or without the space shuttle docked to the station. NASA was incredulous. But the schedule was adhered to and on Saturday morning, April 28th, 2001 Tito and his two crewmates blasted off. There was a lot of tension among the guys at Baikonur towards NASA's possible reaction. Would the Americans refuse to undock the space shuttle? Would Goldin blame the Russians for a screwed up mission and send along a bill for millions of dollars? If so, Koptev would take the blame and the brunt of any political heat from the Kremlin. And it would all, as usual, be the fault of RSC Energia. But we never went down that road. Hours after liftoff, while the Soyuz was still chasing after the ISS, NASA announced that the computer problem had been resolved. Was it a real crisis? We never found out. Soon enough, Semenov's Soyuz docked, as every rocket did under his and Zelenschikov's control, safely, and on schedule.

Tito's time onboard was uneventful from the perspective of trouble being caused or mishaps. His return on May 6th also went off without a hitch. It was like paradise, exclaimed the now sixty year old Tito. "I love space," he had announced on his arrival to the space station and nothing had happened to change that view. The Tito mission was over and the International Space Station had survived. For the first time a passenger paid for his own ticket to space. An American businessman had fulfilled a lifelong dream and opened the excitement of space to millions. For years, many Americans knew the name of Tito while no one, even those in the space industry, could name the NASA astronaut occupying the ISS.

Yet the dispute between NASA and the Russians was not over. Having lost the program battle over Tito, Goldin was now determined to ensure he also lost the war of public opinion. After the mission, the NASA administrator stubbornly blasted Tito to reporters and at a Congressional hearing. He lambasted Tito's supposed ego and sought to lay a huge guilt trip on the first space tourist to ISS. "The current situation has put an incredible stress on the men and women of NASA," Goldin told a House subcommittee. His continued harping made NASA seem bitter and petty. So too when Senator Barbara Mikulski, the fiery Democrat who a decade ago had welcomed Semenov's overture to America, took to the Senate floor and called those who worked on the Tito mission "pimps."

The American establishment weighed in with their own scorn. Senator John Glenn told CNN's *The Capital Gang* how the Tito mission was "a misuse of the spacecraft." *The New York Times* jeered that Dennis Tito had "invented the most offensively elitist form of eco-tourism yet devised by earthlings." Concluded the *Times* editorial, "there is something fundamentally offensive about letting people with a few million to spare piggyback on space vehicles built with billions of dollars of public money."

Ever the chess-player, Semenov concluded that Goldin's bellicose behavior had served a useful purpose. NASA had made a bad mistake, he analyzed. "Everyone now knows that Energia is paying for the Russian obligations out of our own commercial pocket." Semenov was right. At long last, the paradoxical clash between Semenov and Goldin, between the commercial Russian company called Energia and the American organization called NASA, had spilled over into the general public. We went from being the "men in black," those never seen by the public, to having a far more human face.

Semenov seized on the spike in media interest, speaking with European and Russian journalists. In one article with a Russian reporter, Semenov gave a lengthy explanation. Flying commercial, he explained, was critical as "we have to earn money." And once again he tried to explain the dilemma confronting the General Director of the world's most operational space company. "Although the program of Russian participation in ISS development has federal status, its funding is provided at about 50% level. And all this happens in spite of numerous statements about its high priority made at the highest level, for example during a meeting between presidents V.V. Putin and G. Bush.

"But the work on the program is performed on a continuous basis. Each Soyuz TM or Progress vehicle takes almost two years to build. Therefore, in order to be able to send crew and cargo to orbit today or tomorrow, we must procure the necessary materials and place the orders for long-lead items no less than two years in advance. We had and still have to borrow money, take money away from other, commercial programs."

Little had changed since I had first met Semenov. Politicians promised money but no funds were sent. His financial situation was his alone to solve, and using every trick this magician had managed to keep the Mir flying and now the ISS, while juggling the wishes of nationalistic Russian politicians, American misunderstandings and his own desire for a more commercial program.

A few weeks after Tito returned to earth Derechin asked me to compile a sample of the comments from the American press. I sent along a big stack of commentaries, editorials and articles, including, I'm pretty sure, a small piece from the on-line edition of the *National Review*, which was entitled *"Cosmic Capitalism. Tito's take-off may be one small step for free enterprise, but, for the rest of us, it could be a giant leap."* The opinion piece by contributing editor Andrew Stuttaford ended with the exhortation that "if space really is to be opened up, it is going to take more than governments and their "professionals" to do the job. The real work will be done, as it has always been at every new frontier, by the usual motley suspects, by capitalists, cranks, charlatans, and crackpots, by dreamers, drones, visionaries, hucksters, showmen, and opportunists and, yes, even by tourists."

Tito returned to America greeted as a hero by many in the New Space community. I don't believe he received the recognition that he deserved or expected. In 2002 the Boy Scouts of America did bestow on Tito their award for "Americanism." It must have been a moment of vindication for the financial executive hailed as "unpatriotic" by the head of NASA.

Victor Legostaev was not in the least surprised by NASA's grumpy reaction. He reminded me that this was an unfortunate pattern by NASA since the time of Yuri Gagarin's flight, when the first man in space was welcomed by every country in the world except for the United States. Sitting together in his office, Victor Pavlovich provided proof of how long are the memories within the Russian space community. "Even the Queen of England hosted a banquet for our Soviet hero," began the professor. "Befitting the pomp and circumstance of a Royal Dinner, at the head table sat more than a hundred couples and another several hundred in a great hall." Legostaev's eyes sparkled as he retold with apparent pleasure this anecdote

THE FINAL CONFRONTATION

first told to him by Gagarin himself. "In front of our cosmonaut were a half dozen glasses waiting for wine and champagne, and at least a dozen knives, forks and spoons, laid out with military precision. What do you call that, the knives and forks?"

"Silverware?"

"Yes. So much silverware!"

"The Queen noticed that Yuri Alexeyevich was not eating. Why was this?" She asked her guest. "With some embarrassment Yuri Alexeyevich confessed he had never seen so much silverware, and was not sure which fork to use first. This hero, and don't forget Jeffrey, he was such a young man, explained to the Queen that he didn't want to make a mistake in front of all her guests.

"And, and," said Legostaev, finally coming to his point, "the Queen replied, 'Cosmonaut Gagarin, whichever fork you choose to use is the correct fork!'" And with that Legostaev chuckled his whispery laugh reserved for the most ironic of moments.

This anecdote, like all the others that is the oral history of a program falling back into secrecy, will be told and retold as long as Russians journey into space. For me, it was one more opportunity to understand a little more not only about Russia, but also about us-and our own space program as seen through the eyes of others.

As with all of Legostaev's stories, there was a larger purpose than merely retelling a colorful incident. The professor was implying that the public battle over Dennis Tito was not just the doing of Dan Goldin. Organizations, believe men like Legostaev, have their own souls and their own psyche, and embedded within NASA is a nostalgic pride of an earlier time, when the space agency was a beloved symbol of America's Cold War triumphs. This pride has made the transition to the now necessary era of open markets and private competition that much more difficult. But all the more necessary.

Epilogue

What does this all add up to? What did the effort of so many market reformers battling against NASA produce for the American space program? I thought about that a lot after the MirCorp experience, and was unsure of what the final verdict would be. But a chance invitation provided me with the much-needed closure that I was seeking.

In April of 2008 I was invited to attend a symposium in Beijing, China on the future of international relations with the Chinese space program. The event was an eye-opener, as the Chinese showed the same pride, the same determination, the same eagerness to work with NASA and yes, the same conviction that their space program must be commercial that I heard from the Soviets more than two decades before.

Wu Yancheng, a vice president of CASC, which is the rough equal of an earlier Energia organization, confirmed that once further technical progress was realized, the Chinese manned program would be open to foreign partners and customers. "We of course want to work with NASA," one Chinese government official later admitted in Washington. "But that is not of real interest. There is no money there. It is only symbolism."

The Beijing symposium was my return to the space industry after a four year absence. MirCorp did continue after the destruction of the Mir and the flight of Dennis Tito. We represented boy band Lance Bass of 'N Sync, along with "Astronaut Mom" Lori Garver, a former NASA official who is now deputy administrator of the space agency. Both sought to implement a business model using corporate sponsorship as a means to pay for their Soyuz seat. Both failed to raise the necessary level of sponsorship, though Garver was far more successful than Bass in signing corporate clients.

After working with Bass and Garver, the thrill was gone for me. Deciding that a break was definitely required, I went off to write a book on media censorship during the Lincoln Administration. "But that has nothing to do with space," quizzed my friends. Yes, exactly. I then set up a company for innovative space-themed games just outside London. I wanted nothing to do with any project dependent on government space agencies or foreign policy. But from afar I watched with growing interest the stirrings of a new relationship developing between NASA and industry, one that began where we had left off. Eric Anderson's Space Adventures worked with Rosaviakosmos to send paying clients to the ISS space station, without criticism from NASA. In a new generation of space tourism, Microsoft co-founder

Paul Allen supported aerospace engineer Burt Rutan's effort to develop a suborbital vehicle. The duo then combined with Richard Branson and have announced plans to ferry tourists to the start of space. These flights are scheduled to use a spaceport being built by the state of New Mexico, not by NASA. Even better, serious competition has emerged, including one called XCOR.

NASA didn't fight Branson's Virgin Galactic suborbital plans-instead the space agency signed up as a customer. Another American innovator, Robert Bigelow, has announced his intention to develop a commercial space station using discarded NASA technology. Two prototypes of his commercial station have already been launched via Russian rockets. Internet entrepreneurs Jeff Bezos of Amazon and Elon Musk of Paypal have poured their own millions into development of low cost launch vehicles.

From the top, NASA administrator Michael Griffin fought to finally retire the space shuttle fleet, to be replaced with a simpler manned vehicle that emphasizes safety and economy over cutting-edge bells and whistles. Griffin's NASA awarded contracts for private companies to ferry cargo via their own rockets to and from the ISS once the space shuttle is, at long last, retired; and talk of using these private companies for manned transportation has grown louder in volume.

Recognition of the heritage of these developments was expressed in *New Scientist*, which published on September 5th, 2007 a list of the *Top Ten most Influential Thinkers in Space*. Included with Sergei Korolev, Arthur C. Clarke and Tsien Hsue-Shen, was Walt Anderson. Greg Klerkx wrote how "Well before Richard Branson, Paul Allen, Jeff Bezos and other rich entrepreneurs joined the private space game, there was Walt Anderson. A multimillionaire telecommunications mogul and space enthusiast, Anderson bankrolled many early private space ventures and paved the way for the "astropreneurs" who followed.... Anderson's most high-profile space investment was MirCorp.... MirCorp's first client, entrepreneur and former space scientist, Dennis Tito, would go on to become the world's first self-financed space tourist, flying to the International Space Station only a few months after Mir's demise. That event is credited with beginning the private space age in earnest."

My appreciation of this acknowledgment was tempered by the fact that Walt Anderson was imprisoned in a New Jersey penitentiary when the article was published, having been convicted in federal court for tax evasion. It was an emotional blow for many in the New Space industry and a regrettable waste of his talent. There must have been some hesitation on the part of the editors of *New Scientist* to even include Anderson.

But something tugged at me about the tribute, however deserved. Anderson could not have undertaken MirCorp without the active support of Yuri Semenov and his hidden band of, first Soviet, and then Russian supporters. Yet much about Semenov's efforts as well as that of MirCorp and its impact is now being forgotten. In some official space records the MirCorp manned mission is not even included. In Moscow, the pioneering stance of Energia is being paved over by the Putin drive for centralization of all industrial resources back within the Kremlin. It seems a temporary wrinkle in history produced the paradoxical environment allowing a group of Russians to push through the long needed market innovations in space exploration.

Incredibly, I ran into Yuri Semenov at a Chinese restaurant on Prospect Mira in Moscow not so long ago. Forced to retire in 2005 when the Russian Space Agency backed another candidate, the elderly man moved cautiously to the round table near the back of the crowded restaurant. I walked up to Semenov. "Manber?" He bellowed as usual. Questions and recollections came rapid fire. What was I doing? Remember those negotiations in Crystal City with NASA? Remember the times we went swimming in the ocean after the Shuttle launches to our Mir? Then wistfully. "Manber, we were this close." And the old man held up those huge fingers, holding two of them an inch part. "This close."

I asked about Legostaev. He had been named a distinguished member of the Russian Academy of Sciences, a deserved recognition of his long service to his country. And Koptev I had heard was now a powerful government official in the defense community.

Watching the market reforms taking root in America was a wonderful experience, yet the amnesia towards their origins disappointing. I felt a need to fill in the gaps and began writing this book at the start of 2008. After a few months I stopped because of a troubling question. Did Semenov and his team truly embrace free markets, or was it, as so many in America dismissively believed, only the actions of desperate men prepared to do anything to survive? I had wanted to ask that question in the Chinese restaurant, but my Russian wasn't good enough to give it a try.

Beijing settled that question. I realized it just didn't matter why the de facto leader of the Soviet space program embraced commercial markets and economic reforms. Whatever the reason, his stance has transformed the future direction of space exploration, from Washington to Beijing, for the better. Forgetting what transpired, and why, will only slow down the continued reshaping of our space exploration industry.

Where to now?

The market reforms must continue. America's space exploration has become far too expensive, and technology is not the culprit. Industry protectionism drives up the cost and inhibits innovation, resulting in multi-billion dollar weather satellites and space hardware decades late. It's time for NASA and all other government space agencies to play fully by the rules of the international marketplace.

Foreign vendors and private contractors must be allowed to bid to supply all our non-strategic space requirements. Participation into ISS or future planetary programs should require the United States and our trading partners to adhere to a level playing field for procuring hardware, as in aviation. If that means Japanese construction companies building facilities on the lunar surface for U.S. government researchers sponsored by American media organizations, so be it. As Commerce Secretary Baldrige was so fond of saying, space must be just another place to do business; and it is us, in America, who should take the lead in this push for true market reform. The days of a guaranteed all-American or all Russian or all-Japanese or all-European space program must grind to a halt.

But how far we have come. Less and less one hears criticism towards breaking down the barriers keeping out competition and private sector innovation from outer space. Today, whether from the new NASA administrator or a traditional aerospace contractor or Congressional space supporter, it is intuitively accepted that the once-ridiculed pitch of Yuri Semenov is on target.

Competition creates more jobs, not less. Reliance on one launch vehicle or one national program for civil and operational space efforts is wrong. Two is better. Three or four, and pretty soon you have a robust, competitive marketplace. Finally, at long last, I'm confident that's what the future holds for space exploration.

Further Reading

Given that *Selling Peace* is a recollection of personal experiences, much of the information is taken from my own notes. I began publishing my observations and thoughts on the Russian space program soon after I returned from Moscow in December of 1989. My first two articles, one in *Space News* and the other in the Conference Board magazine *Across the Board*, were published within several months. In that regard I was fortunate to have been forced to keep an exact accounting of the earliest events.

But memory and handwriting and 3.5 inch floppy disks do not always make for ideal data sources. Fortunately there is a supporting wealth of information. Bill Broads' articles in *The New York Times* provide a journalistic narrative to the Soviet's first efforts to market their space program to the West. His reporting covers the milestones from the first puzzling overtures by Mikhail Gorbachev's Glavkosmos to the natural conclusion that was the private Russian company RSC Energia. Jeff Lenorovitz, who handled MirCorp's public relations, was for a decade *Aviation Week and Space Technology*'s reporter covering the Soviet space program. Reading both reporters provides a timely accounting of events.

On the Internet, there are two sites that are warehouses of great data. NASA Watch, (www.nasawatch.com,) run by industry observer Keith Cowing, has an archive on key events in the life of MirCorp. In terms of Russian space program hardware, one of the best English language sites is Anatoly Zak's www.russianspaceweb.com, which covers the program from 2001 onwards.

Also available is the NASA Oral History project (http://www.jsc.nasa.gov/history/shuttle-mir.htm), which was invaluable in providing me with renewed insight into the thoughts of senior NASA officials.

Two books capture the role of space exploration on our political scene, and the role of politics on the exploration of space, which are subjects dear to my heart. It just seems we continue not to know where to place space exploration in our overall society. Is it a market? Is it a government project? Is it a jobs program or propaganda tool? William E. Burrows' *This New Ocean* (Random House, 1998) provides a vast sweep of the space race, with an equally comprehensive treatment of the Soviet and the Russians. In *...the Heavens and the Earth* (Basic Books, 1985) Walter A. McDougall digs deeply into the political nature of space exploration and has some of the best material on Kennedy and Eisenhower's differing approaches to space exploration.

Two books provide differing looks on the cooperation between NASA and the Russians and the status of NASA today. *Dragonfly*, (HarperCollins, 1998) provides illumination on NASA's view of working with the Russians, particularly for the seven astronauts who lived on the Mir. Brian Burrough focuses on the infighting between the astronauts and NASA management, and just how unprepared NASA was for cooperation with the Russians. In *Lost in Space* (Pantheon, 2004) Greg Klerkx hits hard at the failures of NASA and offers a very accurate depiction of MirCorp. He earned my respect for his stubbornness in tracking me down in Europe to discuss issues which were then still raw. James Oberg is also an invaluable source of information. The former NASA analyst has written several books, including *Star-Crossed Orbits* (McGraw-Hill, 2002). We disagreed through the early 1990's over the state of affairs of the Russian program, but no American has published more detail on Russian program operations. Also helpful in understanding the cooperation that took place in the 1990's is Ray Williamson's *U.S. Russian-Cooperation in Space* (Office of Technology Assessment, 1995), which gives a snapshot of ongoing joint ventures.

Power and Purpose: U.S. Policy Toward Russia After the Cold War authored by James M. Goldgeier and Michael McFaul (Brookings Institution Press, 2003) offers an excellent depiction of the Clinton Administration's efforts on the space program and specifically their desire to swap the Russian deal on rocket engines with India for a role in the American space station program. *Creating the International Space Station* by David M. Harland and John Catchpole (Springer, 2002) and *The Superpower Odyssey: A Russian Perspective on Space Cooperation* by Yuri Y. Karash (AIAA, 1999) were both helpful in reacquainting me with the order of events during the fast moving time period of the mid-1990's, when the two nations were both eager for an agreement, no matter the lasting consequences.

For those interested in understanding more about the early days of RSC Energia, there is no better English language guide than James Hartford's *Korolev* (Wiley, 1999). Also indispensable is *Rocket and Space Corporation Energia* (Apogee Books, 2001) which is the English edition of the multilingual guide published by RSC Energia. Vladimir Syromiatnikov's *100 Stories on Docking* (Universitetskaya kniga, 2005) is a technical yet personal depiction on key events leading up to Soyuz-Apollo.

FURTHER READING

Nothing matches the Russian-language encyclopedic effort labored over by Yuri Semenov and dozens of Energia archivists for unique information on the Russian space program than *Raketno-Kosmicheskaya Korporatsyia "Energia"* (Energia, 1996). This massive book was the culmination of Semenov's efforts to provide the sort of transparency he thought NASA and the West would require for long-term cooperation.

In a different medium, it's wonderful that Michael Potter has produced a documentary on MirCorp, entitled Orphans of Apollo. Information on availability can be found at http://www.orphansofapollo.com.

Two books which helped provide me with an open mind on working with the Russians should be mentioned. On the political front, civil space cooperation between Russia and America was long a goal for many Democrats in Washington. The most interesting example was the effort of Senator Spark Matsunaga of Hawaii, whose *The Mars Project* (Hill and Wang, 1986) is the ruminations over the merits of pursuing a long term linkage between our two programs. I was introduced to Senator Matsunaga by Harvey Meyerson, who I know played a role in engaging then Senator Al Gore towards the idea of engaging the Soviets.

Finally, I enjoy books that reveal the Russian people in different moments in their long history. One of my favorite captures the mood right after World War II. It is a first person accounting by the great American writer John Steinbeck who, along with photographer Robert Capa, spent several months in the Soviet Union listening, talking to, and drinking with Georgians, Russians and Ukrainians. There is no more accessible window into the Russian people than *The Russian Journal*. Originally published by Viking in 1948, there is now a paperback edition. (Paragon House, 1989).

Space cooperation between two great nations is a never ending effort and the road will undoubtedly veer in unexpected directions during the Obama Administration. I've created www.sellingpeace.com as a means for us to stay in touch. I look forward to hearing from you.

INDEX

2001, A Space Odyssey 296
3M 21
Abbey, George 76, 108, 123, 129, 133, 146, 173
Achenbach, Joel 306
Afanasyev, Viktor 183
Agence France Press (AFP) 12
Akiyama, Toyohiro 71, 277-279
Albrecht, Mark 76, 98, 100
Aldrich, Arnold (Arnie) 6, 81, 83, 93, 97, 101, 104, 106-107, 133, 241
Aldrich, Eleanor
Allen, Joe 27, 65
Allen, Paul 322
Anderman, David 172
Anderson, Eric 301, 321
Anderson, Walt 159-160, 163, 165-169, 171, 175-179, 182-183, 186-189, 191-197, 199-201, 203-205, 208-209, 211, 213, 216, 218, 220, 224-225, 232, 234, 250, 256, 268, 272-275, 277, 283, 289, 291-293, 296, 300, 302, 310-311, 315, 322
Ansari X-Prize 76, 165
Anselmo, Rene 48, 272
APAS Docking System 82, 110, 263
Apollo-Soyuz program 24, 52, 70, 82-84, 106
Ariane 38, 48
Arrott, Anthony 6, 19, 23, 41, 49, 57, 63, 65, 133
Arrott, Anthony S. 33, 47, 49
Artemov, Boris 6, 50, 73, 76-78, 80, 86-87, 131, 133, 141, 304
Assured Crew Rescue Vehicle 93
Astrium 291-292, 299
Astrovision 206-207
Baikonur 49, 53, 57, 156, 182, 233, 255, 260, 315-316
Baklanov, Oleg 141
Baldrige, Malcolm 17, 26, 29, 38, 75, 324
Bass, Lance 12-15, 188, 277, 283, 293, 321
Baturin, Yuri 312
Bella, Ivan 183
Bezos, Jeff 173, 322
Bigelow, Robert 322
Blagov, Victor 109, 233, 258-259, 263-265
Blaha, John 228
Boeing Aerospace 45, 51, 55, 59, 74, 82, 84, 89-91, 104, 108, 112-114, 116, 121-122, 147, 171, 174, 190, 207, 214, 224-225, 233, 242-243, 253, 289
Borenstein, Seth 227-228
Boright, John 91
Botvinko, Alla 189, 221, 256, 259-260, 262
Boy Scouts of America 318
Brainpool 291
Branets, Vladimir 132, 149-150
Branson, Richard 239, 322
Brezhnev, Leonid 143
Broad, William J. (Bill) 35, 40, 136
Brookhaven National Laboratory 64
Brown, George 97, 119
Brumley, Bob 26-32, 34-36, 38-40, 43, 46, 63, 66, 206
Budarin, Nikolai 136
Buran 71, 101, 135, 146, 156
Burnett, Dennis 91
Burnett, Mark 277, 283-287, 289-292, 295, 299-300
Burrough, Brian 306, 326
Bush, George H. W. 25, 45, 49, 72, 76, 97, 103-104, 121, 127, 309, 311-312, 317
Cabana, Robert 312
California Institute of Technology 65
Cameron, James 268-271, 273, 276-277, 285, 288-291, 295, 302
Canadian Space Agency 235, 237-238
Carreau, Mark 232
Carroll, Joe 162-163
Cato Institute 306
Center for Innovative Technology 81
Chafer, Charlie 206
Challenger disaster 42, 53, 85
Chernomyrdin, Victor 102, 107, 116-117, 159, 193, 206
Chertok, Boris 140
Chretien, Jean-Loup 133
Christensen, Sandi 81
Citizen Explorer 239, 267-268, 270-272, 275, 278-280
Clarke, Arthur C. 296, 322
Clinton Administration 115-116, 148, 326
Clinton, Bill 111, 310
Cowing, Keith 247, 325
Cremins, Tom 6, 123, 133
Culbertson, Frank 146, 148-149
Dailey, Brian 97, 100-101, 103-107, 116, 127, 133, 154
Daly, John 185-186, 259, 295
Data Device Corporation (DDC) 149
Day, Dwayne A. 157
DDC 149-151, 153-154
Deep Throatskii 193, 217, 223, 263, 291, 298, 311
Defense Threat Reduction Agency 252
Demisch, Wolfgang 140
Denver, John 12-14, 276
Department 88 24-25, 142
Department of Commerce 25
Department of Defense 31-33
Derechin, Alexander 5, 87-88, 93-94, 99, 103, 108, 111, 118, 128, 130, 133-134, 143, 147-148, 152-153, 155, 159, 171-172, 176, 178, 181, 188, 190-194, 199, 203-204, 207-213, 215-220, 222, 226-227, 232, 252, 258-260, 271, 275, 280, 286-287, 290, 298-301, 310, 313-314, 318
Destination Mir 284-285, 287-288
Destination Space 292
Diamandis, Peter 165
DTRA 252
Dula, Art 54, 64, 74
Duma 72, 78, 142, 148, 153, 179-180, 182, 192-193, 209, 212, 219-220, 223-224, 305
Dunayev, Alexander I. 35, 53-55, 115
Dunstan, Jim 7, 168, 214
Economy, George 19, 39, 41, 44, 65
Eddy, Andrew 235, 237-239, 269, 273, 287, 299
Endeavour computer crash 315
Energia 50th anniversary 5, 139, 141, 143
Energia 5-6, 12-13, 15, 24-25, 50, 52, 55, 59, 61, 64, 69-85, 87-95, 97, 100-104, 106-137, 139-144,

147, 149-156, 161-163, 167-169, 171-173, 176-177, 179-196, 199-203, 205-206, 208-210, 212-223, 225-226, 229-231, 233-234, 238-245, 250-254, 256-259, 261-264, 268, 275-276, 278, 281, 283, 285, 287-289, 291-295, 297-304, 306-307, 310-317, 321, 323, 325-327
Energia Ltd 6, 78, 81-83, 88-89, 110, 117, 126, 133, 150, 152, 217, 233, 262, 312
Esprit Telecom 166
European Space Agency 104, 123, 125, 234, 249, 252, 264, 276
Export Control Administration 31
Faget, Max 27, 65
Faranetta, Christopher (Chris) 6, 63-64, 69, 73, 80, 84, 86, 111, 128, 131, 139, 143, 150, 152, 161, 172
Farber, Gregory 65
Fawkes, Gregg 25-28, 30-31, 34-35, 37, 40-41, 43, 46, 63, 65
FGB (Functional Cargo Block) (Zarya) 116, 119, 148-149, 151
FINDS 161-163, 171
Flade, Klaus-Dietrich 71
Fletcher, James 38
Florida Today 157
Flowers, Ann 19, 32, 37, 42-43, 45, 47, 49, 52, 214
Foale, Michael 146-147
Foundation for the International Non-Governmental Development of Space 161
Fox network 270
Fred Hutchinson Cancer Research Center 65
Freedom 21, 27, 30, 39, 42, 47, 49, 63, 65, 69-70, 82-83, 85, 90-91, 93-94, 97-98, 112, 114, 116, 133, 161, 233, 295, 304
Gagarin, Yuri Alekseyevich 9, 24, 52, 55, 79, 95, 142, 214, 257, 280, 318-319
Gallium arsenide 26
Galtseva, Nellie 6, 218
Gardellini, Gus 6, 161, 165, 171-172, 175-176, 179, 189, 199, 205
Garver, Lori 321
Gascom 153
Gasprom 153
Gidzenko, Yuri 145, 153
Glavkosmos 24, 35, 39, 44, 53-55, 59, 62, 70-71, 74-75, 101, 114-115, 217, 325
Glenn, John 317
GlobalSecurity.Org 13
Globalstar 128
Gold & Appel 213, 221, 225, 296
Golden Apple (The) 232, 296
Goldin, Dan 45, 87, 97-98, 100-101, 103-108, 110-112, 114, 116-117, 119, 121, 123, 135-136, 146, 149, 154, 162, 167, 184, 190, 196, 206-207, 224, 226-230, 232, 238-245, 252, 260, 263-264, 271, 275, 277, 296, 309-311, 314-317, 319
Gorbachev, Mikhail 21, 24, 32, 34, 50, 55, 58-59, 69-71, 75, 141, 174, 325
Gore, Al 102, 107, 116-117, 119, 159, 193, 206, 244, 327
Gore-Chernomyrdin Commission 107, 117
Gorshkov, Leonid 132, 233
Grant, Dick 112, 114
Griffin, Michael 77, 311, 322
Grigoriev, Anton 185
Grigorov, Edward 132
Gromov, Sergei 226-227
Haignere, Jean-Pierre 71, 184

Hanks, Tom 267, 270
Hardman, Helena 298
Harkin, Tom 118-119
Hillary, Edmund 279
Hogan & Hartson 32
Hudgins, Ed 306
Hungry Duck 283
Illuminatus! Trilogy (The) 296
ILS 127
IMAX 130, 269, 276
Institute of Biomedical Problems 147
Intelsat Organization 48
International Launch Services 127
International Space Station 12, 14, 65, 116-117, 119, 145, 148-149, 151, 153-154, 159, 161-162, 181, 185-186, 190, 211, 214, 230, 232, 235, 237, 240, 244-245, 253, 256, 274, 279, 291-292, 300, 306-307, 309, 311, 316, 322, 326
ISS 148, 150-151, 154-155, 161, 186, 190, 221, 229-230, 237, 242-243, 249, 256-257, 261, 268-269, 271-272, 289, 291-293, 298, 300-302, 305-306, 311-318, 321-322, 324
ITAR 32, 43
Jackson-Vanik Amendment 30
Jacobson, John 6, 175-176, 191-192, 195, 204, 211, 213, 215-216, 218, 220, 252
John Deere 21
Johnson and Johnson 21
Johnson Space Center 50-51, 55, 63, 76, 84, 88, 94, 108, 116, 123, 125-126, 149, 152, 180, 312
Juno project 54
Kaleri, Alexandr Yuriyevich 255, 257-258, 262, 275, 279
Kaliningrad (Korolev) 50-51, 56, 75, 77, 90, 118, 135
Kara, Yuri 185
Kathuria, Chirinjeev 6, 210-211, 215, 221, 225, 235, 238, 253, 256, 274, 300
Keller, Samuel (Sam) 81-86, 91-92, 94-95, 97, 100-101, 103-104, 106-107, 109-110, 116, 131, 133, 241
Khrunichev Corporation 74, 117, 127, 202, 241-242
Khrushchev, Nikita 143
Kisselev, Anatoli 74
Klimuk, Petr 184
Kobzon, Yosif 140
Kondakova, Elena 180
Koptev, Yuri 97, 101, 103-109, 114-116, 125, 134, 136, 150, 154, 159, 162, 179, 181, 183-184, 190, 206, 222-224, 226, 228-230, 232, 292, 298-299, 305-306, 309, 312-314, 316, 323
Korolev, Sergei 9, 24, 50, 52, 79, 102, 140-141, 149, 153, 182, 219, 234, 258, 322, 326
Kraselsky, Bruce 41, 165
Krasnov, Alexei 124
Kremlin 24, 30, 44, 59, 70-71, 75, 78-79, 97, 104, 107, 112, 116, 140, 144, 159, 193, 199, 209-211, 217, 220, 258, 298, 301, 304, 316, 323
Krieff, David 293
Krikalev, Sergey 75, 124-126, 132, 136, 142, 145, 153
Kvant-1 61
Kvant-2 61
Last Journey (The) 185
Latyshev, Vsevolod 52, 257, 259-261
Laursen, Eric 73

Lazutkin, Alexander 146
Lebedev, Oleg 132, 233
Legostaev, Victor 9, 12-13, 15, 55, 78-80, 83-84, 86-87, 102, 108-109, 113, 116, 118, 121-122, 128-129, 131, 133-134, 139, 141-143, 146, 148, 152-153, 155, 159, 171, 173, 176, 180-181, 188, 200-201, 203, 226, 238-239, 256-258, 271, 279, 287, 290-291, 310, 318-319, 323
Lenorovitz, Jeffrey 11, 224-225, 227-228, 234, 237, 239, 262, 278, 325
Letterman, Dave 146
Licensintorg 24, 47
Lichtenberg, Byron 19-20, 65
Linenger, Jerry 147, 228, 272, 280
Llewelyn, Peter 184
Lockheed Corporation 45, 51, 69, 76, 84, 127-130, 133, 174, 242, 253
Long-March 3B 148
Loral Corporation 128, 130-131, 148
Los Alamos laboratory 150
Lucid, Shannon 147, 257
Lunar Prospector 73
Macmillan, Mike 268
Makushenko, Yuri 191-193
Maryniak, Gregg 76
McDonnell Douglas corporation 73, 80
McLeod, Dana 41
Men in Black 309, 317
Merscheryakov, Anatoly 47, 50
Microgravity Research Associates 25
Mid Atlantic Telecom 166
Mikulski, Barbara 86, 90-91, 97, 103, 142, 317
Miller, Michael 7, 81
Mir Symposium 126, 258
MirCorp 5-7, 12, 14, 157, 187, 189-193, 197, 199-201, 205-207, 209-210, 212-217, 220-226, 230-232, 234-235, 237-239, 243-245, 249-250, 252-265, 267-278, 280, 283, 285-292, 296-300, 302, 309-310, 321-323, 325-327
Mir-Shuttle 108-109, 115, 117
Mitichkin, Oleg 132
Mitsubishi 174, 289
Mittman, Hank 29, 31, 37, 39, 41, 181
Morring, Frank 6, 111
MTV 293
Muncy, James (Jim) 240
Munitions List 32
Musabayev, Talgat 312
Musarra, Gerald 76, 100-101, 104, 107
Musk, Elon 173, 322
N Sync 12, 14, 277, 283, 321
NASA 5-7, 12-16, 19-51, 53, 55, 59, 61-66, 69-78 79-80, 81, 83, 85, 87, 89, 91, 93, 81-95, 97-109, 111, 113, 115, 117, 111-119, 120, 121-126, 127-137, 139-140-144, 145-146-156, 159-163, 165-169, 171-174, 176-178, 179-188, 190-198, 199-204, 205, 207, 206-210, 211, 221-232, 233, 234-236, 237-246, 247, 249-254, 256-265, 267, 268-281, 287-294, 295-302, 306-307, 309-327
National Space Council 38, 76-77, 98, 100, 103, 112, 123, 127
Nellessen, Wolfgang 133
Nelson, Bill 37, 42-43
Nemstov, Boris 223
New Space entrepreneurs 41, 72, 112, 115-116, 169, 185, 318, 322
Noordwijk 234, 249

Norgay, Tensing 279
NPO Energia 24-25, 50, 55, 69-70, 77, 80-81, 83, 91-93, 100-101, 104, 106-109, 111, 115, 118, 123, 142, 276
O'Brien, Miles 303
O'Connor, Brian 107-108
O'Neill, Gerald K. 63, 76, 82
Oberg, James 45, 326
Office of Space Commerce 26, 28, 39, 46, 48, 65, 81
Office of Space Commercialization 66
OKB-1 24
Oland, Kristan 217, 220, 227, 235, 239, 250, 312
Orbital Sciences Corporation 81
PanAmSat 27, 41, 47-49, 272
Patera, Sasha 57
Payload Systems 6, 19-24, 28-30, 32-33, 35-37, 39-41, 44-45, 48-49, 52, 54, 57, 61-66, 70, 112, 133, 151, 239, 257
Pentagon 29, 32, 34, 37-38, 62, 77, 86-87, 133, 165, 206, 252
Peter the Great 86, 89-90, 135
Pharmaceutical crystal research 32
Pike, John 13-14
Polishchuk, Alexander 304
Pollvogt, Udo 133
Powell, Colin 305-306
Pratt and Whitney 86
Primakov, Yevgeny 217, 223, 230
Progress 53, 56, 58, 62, 71, 116, 146, 162, 173, 183, 185, 187, 190-191, 202, 204, 211, 221, 223, 232-234, 236-237, 242-243, 256, 261-262, 281, 288-289, 301, 314, 318
Proton 27, 54, 61-62, 76, 127, 148, 151, 242
Putin, Vladimir 223-224, 258, 264, 298, 301, 317, 323
Quayle, Dan 98, 105
Radford, Tim 231
Radio Shack 14
Reagan, Ronald 20, 28-30, 34, 38-39, 44-45, 66, 70, 72, 75, 77, 119, 273
Reagan administration 20, 39, 44-45, 70
Regan, Don 38
Reinshaw, Bob 49, 52, 133
Riggs, Conrad 283-286, 288, 291, 293
Riordan, Richard 311
Roberts, Chris 165
Rockwell corporation 44, 49, 51, 59, 69, 73-74, 80, 82, 84, 91, 110
Rohrabacher, Dana 97, 119, 239-243, 245
Rosaviakosmos 150, 153, 162, 183, 217, 219-220, 230, 243, 256, 268, 277, 290, 292, 294, 301, 305, 310, 313, 321
Rotary Rocket 167
RSC Energia 125, 139-140, 142, 144, 149-151, 162-163, 201, 217, 229, 238-239, 257, 289, 291, 306-307, 315-316, 325-326
Russian Space Agency 5, 88, 97, 99, 101, 103, 105, 107-109, 111, 115-117, 123-124, 142, 147, 150, 154, 159, 206, 212, 240, 243, 300, 323
Russian-American relations 174
Rutan, Burt 322
Ryumin, Valery 87, 102, 108-109, 117, 130, 135, 147, 179-181, 187-188, 200, 202, 205, 210, 215, 226, 257, 263-264, 297, 310
Saatchi and Saatchi 54
Salyut 141, 180, 304, 307

Samuels, Jim 25
Sanctis, Edmond 287
Saprykin, Oleg 189
Sawyer, Kathy 35
Schumacher, John 117, 124, 136, 151, 312
Schwartz, Bernard (Bernie) 128, 131, 148
Science Magazine 64
Sea Launch 122, 127, 147-149, 181, 214, 242, 267
Semenov, Yuri 5, 15, 50, 55, 59, 67, 69-76, 78-80, 83-95, 97-109, 110, 112-119, 121-126, 127, 129, 131, 133, 135, 128-137, 139, 140-144, 145-156, 162-163, 167-169, 171-174, 176-178, 179-188, 189-198, 200-204, 205-210, 213-220, 221-232, 238-246, 250-254, 256-265, 271-281, 288-294, 295-296-302, 303-307, 309-319, 323-324, 327
Sensenbrenner, Jim 119, 241, 274
Serebrov, Alexsandr 58
Service Module 116, 149, 190, 243
Sharman, Helen 71, 278-279
Shepherd, Bill 145, 153-154
Shishkin, Oleg 105
Shuttle-Mir 55, 125, 133-134, 146, 179, 187, 241, 325
Shuttleworth, Mark 12, 14
Skylab 116, 183
Slayton, Deke 52, 206
Solovyev, Anatoliy 136
Solovyov, Vladimir 258
Soyuz 12, 24, 52, 62, 70-71, 73, 82-84, 86, 91-93, 95, 106, 113-114, 116, 129, 136, 141, 149, 153, 155, 190-191, 202, 211, 232, 242-243, 255-262, 271-272, 277-281, 283, 288-289, 291-293, 297-298, 300-302, 311-312, 314-316, 318, 321, 326
Soyuz TM 14 71
Soyuz TM-11 278
Soyuz TM-30 255, 257, 277, 281, 289
Space Commander 291
Space Frontier Foundation 163, 169, 172, 289
Space Policy Magazine 25, 41, 45, 76, 161, 225
Space Shuttle 19-21, 23, 25, 33-34, 38, 42-43, 53, 57, 61, 71, 85, 105, 109, 114, 125, 133, 135-136, 146, 161-162, 180, 183, 202, 260, 273, 299, 315-316, 322
Space Station Modules 129, 142, 242
Space Studies Institute 63, 69, 76, 82, 84
Spears, Britney 11-15
Spektr 146
Sputnik 24, 36, 78-79, 95, 168, 214
Stadd, Courtney 76-77
Stafford, Tom 110, 313
Stalin, Josef 24, 57, 143
Star City 52, 61, 147, 184, 257-258, 270, 276, 280, 284, 294, 297, 300, 310, 314
Star Crossed Orbits 45
State Department 27, 32-34, 40, 43, 48, 54-55, 85-86, 90-91, 102, 105, 126, 147-148, 150-154, 159, 189, 196, 202, 206, 230, 251-252, 290
Steklov, Vladimir Alexandrovich 185, 212, 259
Stennis Space Center 206
Stephanopoulos, George 116
Streidel, Brian 6-7, 292

STS-71 135
STS-72 136
STS-84 181
STS-91 180
Syromiatnikov, Vladimir 84-85, 91, 110, 132, 136, 162, 168-169, 189, 262-263, 326
Tether (Firefly Project) 161-163, 171, 173, 177, 189-191, 195, 209, 212, 221, 224-225, 231, 237, 251-252, 254, 263, 269, 290
Thagard, Norman 257
Thomas, Andrew 146, 148
Timberlake, Justin 11-15
Tito, Dennis 12, 14, 267-272, 275-281, 285, 288-289, 291, 296-302, 309-319, 321-322
Tokyo Broadcasting System 278
Top ten complaints 146
Tourists (Space) 15, 40, 77, 92, 239, 250, 255, 279, 300, 302, 318, 322
Travers, Shelly 13
Truly, Richard (Dick) 57, 70, 73, 76, 86, 90, 97, 105, 183, 232, 270, 272, 301-302, 306, 323
Tsibliyev, Vasily 146, 280
Tsien, Hsue-Shen 322
Tsiolkovsky, Konstantin 52
TsUP 47, 49, 52, 55, 61-62, 75, 109, 135, 221, 233, 255-256, 259, 305, 314
Tumlinson, Rick 159, 161-163, 166, 171-173, 175, 177, 179, 187-189, 192, 205-207, 212, 217, 225-228, 231-232, 239, 268-269, 271, 275, 285, 288-289, 302
United Nations Outer Space Treaty 253
United States Space Command 305
Vassiliev, Viacheslav 214-216
Verity, William 29-31, 34, 40, 42, 46
Vibert, Carlo 276
Viktorenko. Alexander 58
Virgin Galactic 239, 322
Vostok 1 79, 142
Walker, Bob 97, 119, 240
Weil, Elizabeth 274-275
Wen Ho Lee 150
Weyers, Gert 235, 250, 262, 283
Wikipedia 137, 279
Wolf, David 67, 146-147, 163
Wu, Yancheng 321
XCOR 322
Yamal satellite 149, 153
Yeltsin, Boris 100, 103-105, 115, 143, 150, 182, 210, 223, 242, 298
Yost, Bruce 49
Yuzhnoy 122
Zalyotin, Sergei Viktorovich 255, 257-258, 260, 262-263, 275, 279, 281
Zelenschikov, Nikolai I. 55, 139, 181-182, 200, 295, 298, 310, 316
Zenit 122
Zhirinovsky, Vladimir 223
Zimmerman, Julianne 49
Znamya 111, 189
Zyuganov, Gennady 223